Cardiac Tissue Engineering

Principles, Materials, and Applications

Synthesis Lectures on Tissue Engineering

Editor
Kyriacos A. Athanasiou and J. Kent Leach, *University of California, Davis*

The Synthesis Lectures on Tissue Engineering series will publish concise books on aspects of a field that holds so much promise for providing solutions to some of the most difficult problems of tissue repair, healing, and regeneration. The field of Tissue Engineering straddles biology, medicine, and engineering, and it is this multi-disciplinary nature that is bound to revolutionize treatments for a plethora of tissue and organ problems. Central to Tissue Engineering is the use of living cells with a variety of biochemical or biophysical stimuli to alter or maximize cellular functions and responses. However, in addition to its therapeutic potentials, this field is making significant strides in providing diagnostic tools.

Each book in the Series will be a self-contained treatise on one subject, authored by leading experts. Books will be approximately 65-125 pages. Topics will include 1) Tissue Engineering knowledge on particular tissues or organs (e.g., articular cartilage, liver, cardiovascular tissue), but also on 2) methodologies and protocols, as well as 3) the main actors in Tissue Engineering paradigms, such as cells, biomolecules, biomaterials, biomechanics, and engineering design. This Series is intended to be the first comprehensive series of books in this exciting area.

Cardiac Tissue Engineering: Principles, Materials, and Applications
Emil Ruvinov, Yulia Sapir, and Smadar Cohen
2012

Central Nervous System Tissue Engineering: Current Considerations and Strategies
Ashley E. Wilkinson, Aleesha M. McCormick, and Nic D. Leipzig
2011

Biologic Foundations for Skeletal Tissue Engineering
Ericka M. Bueno and Julie Glowacki
2011

Regenerative Dentistry
Mona K. Marei
2010

Cardiac Tissue Engineering: Principles, Materials, and Applications
Emil Ruvinov, Yulia Sapir, and Smadar Cohen

ISBN: 978-3-031-01456-7   paperback
ISBN: 978-3-031-02584-6   ebook

DOI: 10.1007/978-3-031-02584-6

A Publication in the Springer series
SYNTHESIS LECTURES ON TISSUE ENGINEERING

Lecture #9
Series Editor: Kyriacos A. Athanasiou and J. Kent Leach, *University of California, Davis*
Series ISSN
Synthesis Lectures on Tissue Engineering
Print 1944-0316   Electronic 1944-0308

Cardiac Tissue Engineering

Principles, Materials, and Applications

Emil Ruvinov, Yulia Sapir, and Smadar Cohen
Ben-Gurion University of Negev
Avram and Stella Goldstein-Goren Department of Biotechnology Engineering

SYNTHESIS LECTURES ON TISSUE ENGINEERING #9

ABSTRACT

Cardiac tissue engineering aims at repairing damaged heart muscle and producing human cardiac tissues for application in drug toxicity studies. This book offers a comprehensive overview of the cardiac tissue engineering strategies, including presenting and discussing the various concepts in use, research directions and applications. Essential basic information on the major components in cardiac tissue engineering, namely cell sources and biomaterials, is firstly presented to the readers, followed by a detailed description of their implementation in different strategies, broadly divided to cellular and acellular ones. In cellular approaches, the biomaterials are used to increase cell retention after implantation or as scaffolds when bioengineering the cardiac patch, *in vitro*. In acellular approaches, the biomaterials are used as ECM replacement for damaged cardiac ECM after MI, or, in combination with growth factors, the biomaterials assume an additional function as a depot for prolonged factor activity for the effective recruitment of repairing cells. The book also presents technological innovations aimed to improve the quality of the cardiac patches, such as bioreactor applications, stimulation patterns and prevascularization.

This book could be of interest not only from an educational perspective (i.e. for graduate students), but also for researchers and medical professionals, to offer them a fresh view on novel and powerful treatment strategies. We hope that the reader will find a broad spectrum of ideas and possibilities described in this book both interesting and convincing.

KEYWORDS

affinity binding, alginate, biomaterials, biomimetic, cardiac tissue engineering, cardiac patches, cardiomyocytes, cell delivery, cell therapy, drug delivery, extracellular matrix, growth factors, heart, hydrogels, heart failure, immunomodulation, liposomes, myocardial infarction, myocardial regeneration, paracrine effect, perfusion bioreactors, scaffolds, stem cells, stimulation, vascularization

*We dedicate this book to our parents
who put their hearts in us.*

Contents

Preface

The diseases of the heart are of a major concern in public health worldwide. The intrinsic inability of the heart to effectively repair itself after major ischemic insults, such as myocardial infarction, urgently calls for continuous extensive research for finding the appropriate strategies to induce myocardial regeneration and repair. While the refined clinical approaches are struggling to improve patient survival, their inability to replace or regenerate a damaged myocardium after myocardial infarction is still a major weakness. Tissue engineering offers novel tools and strategies with a potential to fulfill this need. To help consolidate the developed delicate approaches and to realize the potential of cardiac tissue engineering efforts, this book provides basic knowledge on the various components of the tissue engineering paradigm, and describes the different developed products and outcomes in this field.

Tissue engineering is a multidisciplinary field that combines and utilizes advances in material science, biomedical sciences, and engineering. This book takes the viewpoint of each discipline, and offers findings and references in these areas of research, to present the reader with up-to-date advances in the development of tissue engineering strategies for myocardial regeneration and repair.

Our intent is for the reader to first become familiar with the basic components of the tissue engineering paradigms, namely biomaterials and cells. Various biomaterial and cell types, design criteria, and major issues related to their effect and function are discussed. Next, various choices and strategies using the combinations of those two components are described, where biomaterials can be used only as a delivery vehicle or as a platform for bioengineering of cardiac grafts *in vitro*. In the latter option, the reader will be exposed also to the various strategies aimed to improve mature tissue-like graft formation, by using perfusion bioreactors and various stimulation patterns. In following chapters, realizing the technical and ethical drawbacks of cell transplantation on one hand, and the already established potential on the other, we focus on exciting options for use of acellular forms of biomaterials, with our special interest in injectable, and thus more clinically relevant, forms. And finally, as a significant improvement of this strategy, we give a detailed look at the approaches aimed to induce active myocardial regeneration by employing biomaterial-based delivery of various bioactive molecules, giving a more detailed example of affinity-binding alginate biomaterial.

Educationally, this book can be used a reference book or a textbook for graduate and senior undergraduate students in the programs of biomedical engineering and biotechnology, or help to introduce researchers at various levels to this exciting and continually developing field. For medical professionals in cardiology, this book represents an up-to-date introduction and summary of promising cardiac tissue engineering approaches, offering powerful alternatives or complimentary strategies to currently available clinical therapies.

Finally, we hope to convince all readers to join us in this fascinating journey.

Emil Ruvinov, Yulia Sapir, and Smadar Cohen
July 2012

Acknowledgments

We heartily thank the past and present members of the Cohen group for putting their hearts and souls into advancing the research of biomaterials and cardiac tissue engineering.

We also acknowledge the generous funding provided by

1. Ben-Gurion University- Applied Fund of BGN Technologies

2. BioLineRx Innovation, Jerusalem

3. European Union FWP7 (INELPY)

4. Israel Ministry of Health

5. Israel Ministry of Science

6. Israel Science Foundation (grants #: 52/99-1, 793/04, 1368/08)

7. Fellowships by Daniel Falkner (zl) and his daughter Ms. Ann Berger (United Kingdom)

8. Azrieli Fellowship to Yulia Sapir for PhD studies

9. SC holds the Claire and Harold Oshry Chair in Biotechnology

Emil Ruvinov, Yulia Sapir, and Smadar Cohen
July 2012

CHAPTER 1

Introduction

CHAPTER SUMMARY

The introductory chapter opens with an overview on the current status of cardiac tissue engineering and describes the four strategies evolved in the last decade or so. The objectives and scope of this book are then presented, followed by the monograph organization and short summary of the ten chapters' contents.

1.1 CARDIAC TISSUE ENGINEERING

Tissue engineering is a fast growing scientific and technological field that aims at the repair or replacement of damaged and dysfunctional tissues [1]. This therapeutic strategy is of particular interest for the treatment of heart diseases, where large portions of functional tissue are lost (i.e., after myocardial infarction (MI)) with very limited intrinsic regeneration ability of the heart. Such loss often leads to significant and progressive deterioration in cardiac contractility and function, resulting eventually in the development of life-threatening conditions and congestive heart failure. The shortage of transplants and donors, and the inability of current clinical therapies to restore functional myocardial tissue loss after MI add additional concerns to the already high burden of heart disease in the public health.

In general, the tissue engineering paradigm relies on the implementation of various combinations of biomaterials/scaffolds, cells, and bioactive molecules [2, 3, 4]. In the last decade or so, four main strategies were developed under the concept of cardiac tissue engineering. One strategy uses biomaterials as vehicles for cell delivery and retention in the infarcted heart [5, 6, 7]. The biomaterial vehicle (usually in the form of three-dimensional porous degradable scaffolds), while providing mechanical support for the infarcted tissue, creates a favorable microenvironment for promoting transplanted cell survival and long-term action. A second strategy is the *in vitro* bio-engineering of cardiac patches [4, 8]. Here, biomaterials in various forms (porous solid scaffolds, sheet-like structures, and hydrogels) are seeded with cardiac cells; the cell constructs are cultivated in perfusion bioreactors and under physical stimulation to promote tissue development and maturation, and when this is achieved they are implanted on damaged myocardial tissue for replacement and/or support. The third strategy focuses on the use of biomaterials in acellular forms as structural restrainers and scar filling to attenuate heart remodeling and dysfunction [9, 10]. These strategies are of a particular interest because of the well-defined and characterized nature of the end products, their off-the-shelf availability, the relatively low cost compared to stem cell-based products, and their greater clinical applicability due to minimally invasive procedures applied in their delivery. As a significant advancement of strategy number 3 and to achieve an active cardiac tissue regeneration, the fourth strategy offers combination of biomaterials with bioactive molecules, in the form of local and controlled delivery systems [11, 12]. Such combinations offer mechanical support while simultaneously providing bioactive components, to achieve long-term functional improvements by induction of myocardial regeneration and/or effective tissue salvage.

Progress in cardiac tissue engineering relies on a synergistic effort from a variety of areas of materials science, bioengineering, biomedicine, and cell biology. For example, recent advances in the field of pluripotent stem cells and reprogramming offer previously unavailable sources of human cardiomyocytes for clinical applications [13]. Engineering efforts led to development of perfusion bioreactors and different stimulation patterns, indispensable now for the optimal growth of *in vitro*-generated myocardial patches [14, 15]. Careful analysis of cell biology and the native cellular microenvironment allows for the implementation of mimicking features from native ECM in the design of the optimal cardiac patch [16, 17, 18]. With the advancement in nanotechnology and nano-fabrication techniques, the microenvironment can be designed on the nano-scale, such as with surface patterning, anisotropy, etc. [19, 20, 21].

Cardiac tissue engineering still has several challenges to solve, in order to maximize its therapeutic effects on myocardial repair and regeneration [4, 22]. In current cellular tissue engineering modalities under investigation, the choice of cells is central. Many efforts are and will be invested in preparation of human sources of cardiac progenitors or cardiomyocytes. Endothelial cells of human origin are also required to recapitulate more closely the cardiac environment and promote vascularization. The lack of vascularization in the thick tissue grafts prepared *in vitro* represents another challenge. Various pre-vascularization or *in situ* vascularization strategies (either cell- or bioactive molecule-based) have to be employed to maintain blood supply to the implanted construct. The properties of biomaterials and of the resulting artificial matrix are also of great importance, as cells respond to their environment in very different ways. The mechanical properties and structure of the matrix must match those of native cardiac tissue. A hallmark of functional myocardium is its ability to propagate electrical impulses and to respond to these impulses by synchronized contractions; thus the engineered cardiac construct should have the appropriate electrical properties (propagation velocities and directions, excitation thresholds, etc.) to integrate electrically into existing electrical syncytium of the existing myocardium without causing arrhythmia. Finally, the development of more sophisticated tissue engineering solutions will require development of advanced culture platforms (i.e., perfusion bioreactors, etc.) that also will be able to integrate multiple stimulation patterns and signals.

In the acellular biomaterial-based strategies, although their therapeutic benefits in cardiac repair after MI have been established in animal trials, clinical trials are needed to establish the long-lasting effects of biomaterials. In this respect, identification of the appropriate cocktail of bioactive molecules for inducing cardiac tissue regeneration would be a breakthrough.

Solving these challenges would probably facilitate the translation of the developed modalities into clinics, eventually introducing cardiac tissue engineering as a standard, but state-of-the-art, treatment option for patients.

1.2   OBJECTIVES AND SCOPES

In this monograph, we intend to present an up-to-date view on cardiac tissue engineering, in order to provide students, researchers, and medical professionals with essential information regarding these advanced and effective strategies for myocardial repair and regeneration, to be used either as standalone products, or as combinations with current clinical therapies.

After providing essential information on the major components in cardiac tissue engineering, cell sources, and biomaterials, the different strategies for cardiac tissue engineering are presented. These strategies can be divided into cellular and acellular, while they share a common theme, the biomaterials. In cellular approaches, the biomaterials are used to increase cell retention after implantation or as scaffolds when bioengineering the cardiac patch, *in vitro*. In acellular approaches, the biomaterials are used as ECM replacement for damaged cardiac ECM after MI, or, in combination with growth factors, the biomaterials assume an additional function as a depot for prolonged factor activity.

Our main goal here is to introduce the basic principles in each strategy, discuss the challenges, and offer solutions from our viewpoint. The progress made in recent years including in clinical trials, indicate that cardiac tissue engineering may be a feasible treatment option for diseases of the heart. We hope to convince our readers to join us in this fascinating journey.

1.3   ORGANIZATION OF THE MONOGRAPH

The monograph consists of ten chapters. Chapter 1 gives a short overview of the field of cardiac tissue engineering, its main challenges, and then information on the objectives and organization of the book.

Chapter 2 aims to familiarize the reader with heart physiology, the heart muscle structure, and major diseases of the heart for which cardiac tissue engineering may be the ideal treatment solution (myocardial infarction (MI) and heart defects). The limited ability of the heart to regenerate itself after damage calls for alternative therapies, and the possible therapeutic targets and strategies to induce cardiac regeneration after MI will be presented.

Chapter 3 summarizes the characteristics of the potential cell components in cardiac tissue engineering with an emphasis on stem cells as the most available source. Embryonic, adult, as well as the recently developed induced pluripotent stem cells are discussed while presenting the challenges and their advantages/drawbacks. Finally, clinical trials of injecting cell suspensions are described while presenting the theory on paracrine effects of the cells on cardiac repair.

Chapter 4 deals with another major component of the cardiac tissue engineering paradigm, biomaterials. From our perspective, biomaterials are the most important component in all of the strategies for cardiac repair. Biomaterials treatment of MI has been recently established as a standalone therapy, as will be further described in Chapter 9 for alginate biomaterial, the first-in-man trial of intracoronary delivery of a biomaterial to support a failing heart after MI. Chapter 4 thus provides a brief summary of the "need to know" about biomaterials; such as the basic criteria for material selection and design, the type of natural and synthetic polymers in use, scaffold types, and their fabrication methodology.

From Chapter 5 and on, the design and application of biomaterials in various strategies of cardiac tissue engineering are presented. In Chapter 5, biomaterials are applied combined with transplanted cells to increase cell retention, survival, and function at infarct zone; this information is presented for each cell type investigated in animals trials. Finally, the recent results of the MAGNUM (Myocardial Assistance by Grafting a New Bioartificial Upgraded Myocardium) phase I clinical trial, provide proof-of-concept for the beneficial effects of biomaterials as cell supporters.

Chapters 6, 7, and 8 describe the different aspects in the development of a cardiac cell patch composed of seeded cells within polymeric scaffolds. In Chapter 6, the basics of the patch engineering strategy is presented, with an emphasis on the design of biomaterials and scaffolds including the implementation of micro- and nano-technologies to obtain scaffolds promoting its vascularization and anisotropic cardiac tissue structure. Chapter 7 deals with mass transport in 3D cell cultures and introduces perfusion bioreactors as a solution. Different stimulation patterns to induce cardiac tissue arrangement and maturation within the patch are described, including mechanical, electrical and magnetic stimulation. In Chapter 8, two main approaches for patch vascularization prior to its transplantation on the infarct are presented, the triculture patch with endothelial cells and using the body as bioreactor for pre-vascularization of the patch.

Chapters 9 and 10 introduce an important breakthrough of the last decade or so, the acellular treatment of the heart after MI using acellular biomaterials. This strategy has been especially successful for treating MI, as excessive damage to cardiac ECM is a key effector leading to structural changes in the heart after MI and congestive heart failure. The use of various forms and types of biomaterials as ECM replacements and a substrate for cell infiltration and integration are discussed (implantable scaffolds, decellularized ECMs, injectable polymer solutions, and hydrogels as well as immunomodulating liposomes). In Chapter 9, we describe an alginate biomaterial developed by our group which currently is in advanced clinical trials. Due to its properties, the biomaterial can be delivered via the intracoronary artery to the infarct and only at the infarct to undergo solidification, thus mechanically supporting the heart.

The closing Chapter 10 upgrades the acellular biomaterial strategy by combining the biomaterials described in Chapter 9 and bioactive molecules to promote active tissue regeneration. The chapter presents the motivation for this strategy to attain a long-lasting effect on cardiac repair, and describes the different materials, growth factor combinations, and the approaches implemented to achieve prolonged release and activity of the factors. We then focus on the design of affinity-binding alginate, synthesized by sulfation of the uronic acids, to mimic factor interactions with heparin/heparan sulfate. Our investigations on this novel material are introduced, including proving its efficacy for controlled multiple factor delivery and as a treatment for inducing active regeneration in various diseases and MI.

BIBLIOGRAPHY

[1] Langer R, Vacanti JP. Tissue engineering. Science. 1993;260:920–6. 1

[2] Ruvinov E, Dvir T, Leor J, Cohen S. Myocardial repair: from salvage to tissue reconstruction. Expert Rev Cardiovasc Ther. 2008;6:669–86. 1

[3] Rane AA, Christman KL. Biomaterials for the treatment of myocardial infarction a 5-year update. J Am Coll Cardiol. 2011;58:2615–29. DOI: 10.1586/14779072.6.5.669 1

[4] Vunjak-Novakovic G, Lui KO, Tandon N, Chien KR. Bioengineering heart muscle: a paradigm for regenerative medicine. Annu Rev Biomed Eng. 2011;13:245–67. DOI: 10.1146/annurev-bioeng-071910-124701 1, 2

[5] Nunes SS, Song H, Chiang CK, Radisic M. Stem Cell-Based Cardiac Tissue Engineering. J Cardiovasc Transl Res. 2011. DOI: 10.1007/s12265-011-9307-x 1

[6] Martinez EC, Kofidis T. Adult stem cells for cardiac tissue engineering. J Mol Cell Cardiol. 2011;50:312–9. DOI: 10.1016/j.yjmcc.2010.08.009 1

[7] Segers VF, Lee RT. Biomaterials to enhance stem cell function in the heart. Circ Res. 2011;109:910–22. DOI: 10.1161/CIRCRESAHA.111.249052 1

[8] Ye KY, Black LD, 3rd. Strategies for Tissue Engineering Cardiac Constructs to Affect Functional Repair Following Myocardial Infarction. J Cardiovasc Transl Res. 2011. DOI: 10.1007/s12265-011-9303-1 1

[9] Tous E, Purcell B, Ifkovits JL, Burdick JA. Injectable acellular hydrogels for cardiac repair. J Cardiovasc Transl Res. 2011;4:528–42. DOI: 10.1007/s12265-011-9291-1 1

[10] Singelyn JM, Christman KL. Injectable materials for the treatment of myocardial infarction and heart failure: the promise of decellularized matrices. J Cardiovasc Transl Res. 2010;3:478–86. DOI: 10.1016/j.jacc.2011.10.888 1

[11] Segers VF, Lee RT. Local delivery of proteins and the use of self-assembling peptides. Drug discovery today. 2007;12:561–8. DOI: 10.1016/j.drudis.2007.05.003 2

[12] Ruvinov E, Harel-Adar T, Cohen S. Bioengineering the infarcted heart by applying bio-inspired materials. J Cardiovasc Transl Res. 2011;4:559–74. DOI: 10.1007/s12265-011-9288-9 2

6 BIBLIOGRAPHY

[13] Burridge PW, Keller G, Gold JD, Wu JC. Production of de novo cardiomyocytes: human pluripotent stem cell differentiation and direct reprogramming. Cell Stem Cell. 2012;10:16–28. DOI: 10.1016/j.stem.2011.12.013 2

[14] Radisic M, Marsano A, Maidhof R, Wang Y, Vunjak-Novakovic G. Cardiac tissue engineering using perfusion bioreactor systems. Nat Protoc. 2008;3:719–38. DOI: 10.1038/nprot.2008.40 2

[15] Shachar M, Cohen S. Cardiac tissue engineering, Ex-vivo: Design principles in biomaterials and bioreactors. Heart Fail Rev. 2003;8:271–6. DOI: 10.1023/A:1024729919743 2

[16] Lutolf MP, Gilbert PM, Blau HM. Designing materials to direct stem-cell fate. Nature. 2009;462:433–41. DOI: 10.1038/nature08602 2

[17] Ma PX. Biomimetic materials for tissue engineering. Adv Drug Deliv Rev. 2008;60:184–98. DOI: 10.1016/S0142-9612(03)00339-9 2

[18] Davis ME, Hsieh PC, Grodzinsky AJ, Lee RT. Custom design of the cardiac microenvironment with biomaterials. Circ Res. 2005;97:8–15. DOI: 10.1161/01.RES.0000173376.39447.01 2

[19] Zhang B, Xiao Y, Hsieh A, Thavandiran N, Radisic M. Micro- and nanotechnology in cardiovascular tissue engineering. Nanotechnology. 2011;22:494003.
DOI: 10.1088/0957-4484/22/49/494003 2

[20] Dvir T, Timko BP, Kohane DS, Langer R. Nanotechnological strategies for engineering complex tissues. Nature nanotechnology. 2011;6:13–22. DOI: 10.1038/nnano.2010.246 2

[21] Kelleher CM, Vacanti JP. Engineering extracellular matrix through nanotechnology. J R Soc Interface. 2010;7 Suppl 6:S717–29. DOI: 10.1098/rsif.2010.0345.focus 2

[22] Vunjak-Novakovic G, Tandon N, Godier A, Maidhof R, Marsano A, Martens TP, et al. Challenges in cardiac tissue engineering. Tissue Eng Part B Rev. 2010;16:169–87.
DOI: 10.1089/ten.teb.2009.0352 2

CHAPTER 2

The Heart—Structure, Cardiovascular Diseases, and Regeneration

CHAPTER SUMMARY

A major task of cardiac tissue engineering is to provide functional myocardial tissues for replacement of a damaged myocardium. The engineered tissue features have to mimic the natural myocardial tissue to ensure its successful integration, regeneration, and functional recovery in a reasonable amount of time. This chapter presents a detailed description of the heart and its muscle structure to reveal the complexity with which the cardiac tissue engineer has to cope in the process. Then, two major pathophysiological conditions of the heart, myocardial infarction and congenital heart defects, are introduced, for which cardiac tissue engineering potentially can be an ideal solution. Finally, we discuss the regeneration capability of the adult heart and introduce the targets and possible therapeutic strategies and interventions aimed at inducing myocardial regeneration.

2.1   INTRODUCTION

The heart is a central organ in the circulatory system, and is indispensable for normal organism homeostasis by providing a constant supply of blood to tissues, which carries oxygen and nutrients and removes carbon dioxide and waste products. The heart pumps blood through blood vessels, and to accomplish this, the heart beats about 100,000 times every day, which adds up to 35 million beats in a year and about 2.5 billion times in an average lifetime, pumping 5 liters of blood each minute. The unique heart structure ensures a continuous blood supply, at rest and in various stressful conditions. Due to the central function of the heart in sustaining life and normal homeostasis of the body, the diseases of the heart are a major concern in public health. The insufficient ability of the heart to self-regenerate after damage, such as after myocardial infarction, results in a progressive deterioration in heart function. During the course of this process, the heart undergoes significant structural and biochemical changes, leading to reduced contractility and heart failure. There is no cure for myocardial infarction, and the goal of cardiac tissue engineering is to provide solutions for this devastating disease.

2.2   THE HEART AND CARDIAC MUSCLE STRUCTURE

The heart has four chambers (Fig. 2.1A-B). The two superior receiving chambers are the atria, and the two inferior pumping chambers are the ventricles. The left ventricle of the heart pumps oxygenated blood into the systemic circulation to all tissues of the body except the air sacs (alveoli) of the lungs. The right ventricle of the heart pumps deoxygenated blood into the pulmonary circulation to the alveoli of the lungs.

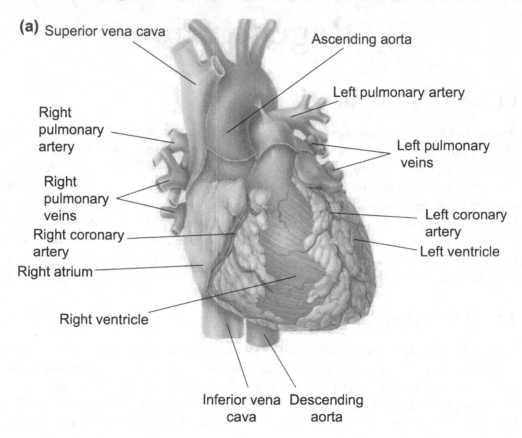

Figure 2.1: Heart and cardiac muscle structure. **A.** Anterior external view of the heart showing major surface features. Reprinted with permission of John Wiley & Sons, Inc. [1].

The wall of the heart consists of three layers: the epicardium (external layer), the myocardium (middle layer), and the endocardium (inner layer). Epicardium, the thin, transparent outer layer of the heart wall, is composed of mesothelium and delicate connective tissue that imparts a smooth, slippery texture to the outermost surface of the heart. The middle myocardium, which is the cardiac muscle tissue, constitutes about 95% of the heart mass and is responsible for its pumping action. The

(b)

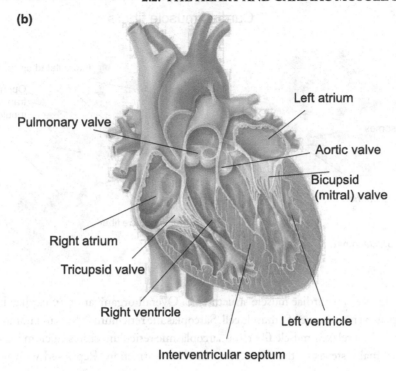

Figure 2.1: Heart and cardiac muscle structure. **B.** Major internal features of the heart. Blood vessels that carry oxygenated blood are colored red, whereas those that carry deoxygenated blood are colored blue. Reprinted with permission of John Wiley & Sons, Inc. [1].

cardiac muscle fibers swirl diagonally around the heart in bundles. The innermost endocardium is a thin layer of endothelium overlying a thin layer of connective tissue. It provides a smooth lining for the chambers of the heart and covers the valves of the heart. The endocardium is continuous with the endothelial lining of the large blood vessels attached to the heart, and it minimizes surface friction as blood passes through the heart and blood vessels.

The human myocardium consists of 2 to 3 billion cardiomyocytes (75% by volume, ~30% by number), the striated muscle cells found only in the heart that can be distinguished from the skeletal and smooth muscle cells. Apart from muscle cells, the heart tissue is mainly composed of fibroblasts (about two thirds in terms of numbers) and endothelial cells. Unlike skeletal muscle cells, cardiomyocytes are controlled by the autonomic (involuntary) rather than the somatic (voluntary) nervous system. Furthermore, cardiomyocytes can generate their own excitatory impulses, functioning as a biological pacemaker.

The myocardium assumes a unique structure, enabling it to synchronously contract (Fig. 2.1C-D). Compared with skeletal muscle fibers, the cardiac muscle fibers are shorter in length and less

(c)   Cardiac muscle fibers

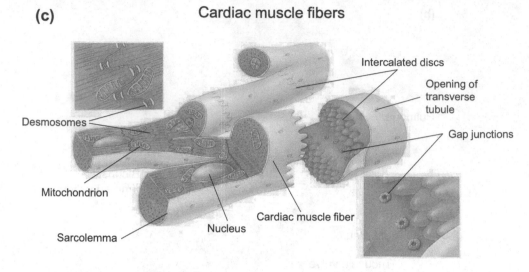

Figure 2.1: Heart and cardiac muscle structure. **C.** Overall organization of cardiac muscle fibers. Sarcolemma is plasma membrane of a muscle cell. Sarcoplasmic reticulum (SR) is membranous sacs encircling each myofibril. In a relaxed muscle fiber, the sarcoplasmic reticulum stores calcium ions. Release of Ca^{2+} from the terminal cisterns of the SR triggers muscle contraction. Reprinted with permission of John Wiley & Sons, Inc. [1].

circular in the transverse section. They also exhibit branching, which gives the individual cardiac muscle fibers a "stair-step" appearance. A typical cardiac muscle fiber is 50–100 μm long and has a diameter of about 14 μm. The cardiomyocyte has one centrally located nucleus, although an occasional cell may have two nuclei. The ends of cardiac muscle fibers connect to neighboring fibers by irregular transverse thickenings of the sarcolemma called intercalated discs. The discs contain desmosomes, which hold the fibers together, and gap junctions, which allow the muscle action potentials to conduct from one muscle fiber to its neighbors. Gap junctions allow the entire myocardium of the atria or the ventricles to contract as a single, coordinated unit.

Myofibrils, the contractile structure of cardiomyocytes, are composed of repeating single contractile units known as sarcomeres (Fig. 2.1D). Electrical excitation of cardiomyocytes leads to contraction of the heart through the process of excitation-contraction coupling (ECC). The ubiquitous second messenger, Ca^{2+}, is essential for cardiac electrical activity and is the direct activator of the myofilaments, which cause contraction [2]. Myocyte mishandling of Ca^{2+} is a central cause of both contractile dysfunction and arrhythmias in pathophysiological conditions [2]. The cardiomyocyte contraction machinery is based on two main proteins, myosin and actin, that build thick and thin filaments, respectively. During muscle contraction, actin fibers move toward the inner space of the sarcomere by sliding along the fixed myosin fibers. Each sarcomere is bounded by Z-lines formed by

(d)   Arrangement of components in a cardiac muscle fiber

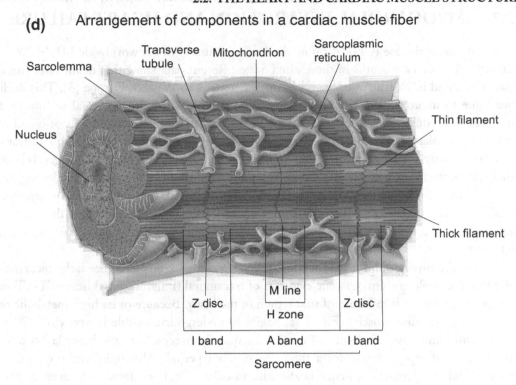

Figure 2.1: Heart and cardiac muscle structure. **D.** Internal arrangement of the cardiac fiber showing basic sarcomere structure. The assembly of contractile proteins into sarcomeres is a complex process that requires coordinate synthesis of the constituent proteins, the polymerization of actin and myosin (and many associated proteins) into thin and thick filaments, respectively, and the association of the two filament systems into highly organized sarcomeres. Newly assembled sarcomeres consist of parallel arrays of ~1.0 μm-long thin filaments that interdigitate with laterally aligned 1.6 μm-long thick filaments. Narrow, plate-shaped regions of dense protein material called Z discs separate one sarcomere from the next. Thus, a sarcomere extends from one Z disc to the next Z disc. The thick and thin filaments overlap one another to a greater or lesser extent, depending on whether the muscle is contracted, relaxed, or stretched. The pattern of their overlap, consisting of a variety of zones and bands (I, A, and H), creates the striations that can be seen both in single myofibrils and in whole muscle fibers. The M line marks the middle of the sarcomere. Reprinted with permission of John Wiley & Sons, Inc. [1].

protein aggregates, situated at the edge of the sarcomere. Together, the protein complexes comprising the sarcomere enable the macroscopic movement associated with contractile activity (Fig. 2.1D) [3].

2.3   MYOCARDIAL INFARCTION AND HEART FAILURE

Coronary heart disease (CHD) is now the leading cause of death worldwide [4]. In 2008, CHD caused ~1 of every 6 deaths in the United States. Recent data show that death rates from CHD have decreased in North America and in many countries in Western Europe [5]. This decline has been due to improved prevention, diagnosis, and treatment (pharmacological or interventional), in particular reduced cigarette smoking among adults, and lower average levels of blood pressure and blood cholesterol. However, the burden of CHD is increasing in developing and transitional countries, partly as a result of increasing longevity, urbanization, and lifestyle changes. It is expected that 82% of the future increase in coronary heart disease mortality will occur in developing countries.

Myocardial infarction (MI) is the most common manifestation of CHD, accounting for ~50% of all CHD cases in the US. MI results from temporary or permanent occlusion of the main coronary arteries (Fig. 2.2), causing significant blood supply reduction to the beating heart muscle (mainly left ventricle).

All the myocardium that is supplied by the occluded artery becomes ischemic, resulting in chest pain and electrocardiographic evidence of transmural (full-thickness) ischemia (ST-segment elevation) in the leads reflective of that region of the heart. Because of its high metabolic rate, the myocardium (cardiac muscle) (Fig. 2.3a) begins to undergo irreversible injury within 20 minutes of ischemia, and a wavefront of cell death subsequently sweeps from the inner layers toward the outer layers of myocardium over a three- to six-hour period. Although cardiomyocytes are the most vulnerable population, ischemia also kills vascular cells, fibroblasts, and nerves in the tissue. Myocardial cell necrosis (Fig. 2.3b) elicits a vigorous inflammatory response. Hundreds of millions of marrow-derived leukocytes, initially composed of neutrophils and later of macrophages, enter the infarct. The macrophages phagocytose the necrotic cell debris and likely direct the subsequent phases of wound healing. Concomitant with removal of the dead tissue, a hydrophilic provisional wound repair tissue rich in proliferating fibroblasts and endothelial cells – termed granulation tissue (Fig. 2.3c) – invades the infarct zone from the surrounding tissue. Over time, granulation tissue remodels to form a densely collagenous scar tissue (Fig. 2.3d). In most human infarcts, this repair process requires two months to complete. Infarcts in smaller experimental animals such as mice or rats heal substantially faster [7].

The sudden loss of a significant portion of myocardium leads to a decrease in contractile function of the heart. These changes are compensated by an increase in left-ventricular volume, which augments contractility in the non-infarcted myocardium via the Frank-Starling mechanism (Fig. 2.2). A negative consequence of the enhanced left-ventricular volume, however, is the intensified stress on the ventricular wall. This is partially counteracted by scar formation in the infarct zone and cardiomyocyte hypertrophy in the non-infarcted myocardium (Fig. 2.2). These mechanisms provide temporary compensation for the loss of myocardium contractility. However, in large infarcts, these mechanisms fail, and further deterioration in cardiac function occurs. In these patients, MI results in thinning of the injured wall and dilation of the ventricular cavity, a process termed ventricular remodeling (Fig. 2.2 and Fig. 2.3e). These structural changes markedly increase the mechanical

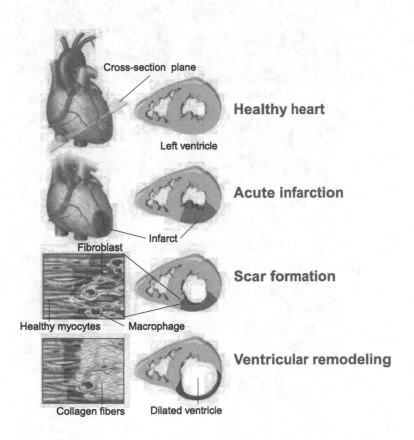

Figure 2.2: Heart failure: from acute crisis to chronic disease. In a healthy heart, the heart's left ventricle pumps newly oxygenated blood to the rest of the body, and its walls are normally thick with cardiac muscle fibers. When blood supply to the beating muscle is reduced as a result of coronary artery occlusion, myocytes die from oxygen deprivation, and the infarct develops. Within hours and days, existing extracellular matrix degradation takes place. The infarct is infiltrated by macrophages and collagen-producing myofibroblasts. The infarcted wall of the ventricle becomes thin and rigid. As healthy myocytes die at the border of the scarred area, the infarct expansion continues. Developed pressure overload is initially compensated by the hypertrophy of the healthy myocardium. Ultimately, however, this compensating mechanism fails, and infarct wall thinning and ventricle dilatation continues. As a result, the heart is unable to pump effectively, leading to life-threatening condition of heart failure. Reprinted with permission from [6].

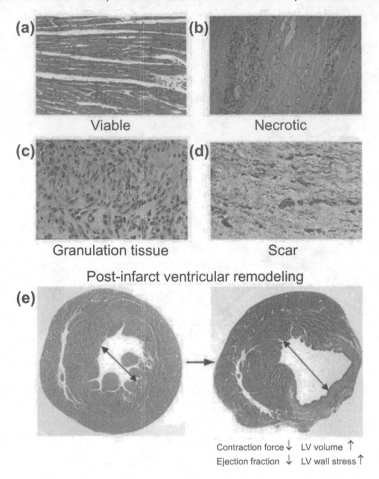

Figure 2.3: Histological stages of myocardial infarction. a-d: changes in infarct area after ischemic event. e. ventricular remodeling. See text for details. Reprinted with permission from [7].

stress on the ventricular wall and promote progressive contractile dysfunction, eventually leading to chronic and congestive heart failure (CHF) [7, 8, 9, 10].

Over recent decades, major improvements have been realized in the management of patients with MI [11, 12]. These include in-hospital treatments (e.g., pharmacological lysis and anti-platelet and anti-thrombin therapies), interventional therapies and surgery (e.g., cardiac catheterization, percutaneous coronary intervention, coronary artery bypass surgery, and heart transplantation), as well as drug regimens for prevention and long-term treatment (e.g., aspirin, ACE inhibitors, β-blockers, and statins). The improved management of acute coronary events, however, has led to a significant increase in the number of patients who suffer from chronic conditions, namely CHF.

A major disadvantage of the above therapies is their inability to replace, at least partially, cardiac muscle loss after infarction. Thus, there is a need for alternative approaches able to overcome the limitations of standard therapies. The ultimate goal of such novel therapies is the induction of myocardial tissue regeneration (therapeutic, endogenous, or combined) *in situ* or *ex vivo*.

2.4 CARDIAC EXTRACELLULAR MATRIX (ECM)—ITS FUNCTION AND PATHOLOGICAL CHANGES AFTER MI

The architectural complexity of the myocardium and the potential role of ECM in maintaining the unique myocyte orientations throughout the LV free wall were described by Streeter and Basset [13]. Using a structural engineering approach, these authors demonstrated that myocyte orientation and myocardial fiber angles are highly organized and move in a continuous fashion from the endocardium to the epicardium. It is the structural network of matrix proteins composed of proteins of highly organized structure and architecture, such as type I and type III collagen, that provide structural integrity to adjoining myocytes and contribute to overall LV pump function through the coordination of myocyte shortening. Scanning electron microscopy studies demonstrated the three-dimensional structure of the myocardial ECM and how the fibrillar weave surrounded and supported individual myocytes as well as fascicles of myocytes [14]. Moreover, these initial studies demonstrated the complexity of the ECM and the structural interaction with the vascular compartment. Further research demonstrated that the myocardial ECM maintains alignment of myofibrils within the myocyte through a collagen-integrin-cytoskeleton-myofibril relation (Fig. 2.4). The main components of myocardial ECM are listed in Table 2.1.

In addition to a fibrillar collagen network, a basement membrane, proteoglycans, and glycosaminoglycans, the myocardial ECM contains a large reservoir of bioactive molecules [14]. For example, it has been demonstrated that the concentration of bioactive signaling molecules such as angiotensin II (ANG II) and endothelin (ET)-1 are over 100-fold higher within the myocardial interstitium than in plasma [18, 19]. Moreover, cytokine activation and signaling such as that for tumor necrosis factor-α (TNF-α) is highly compartmentalized within the myocardial interstitium [20]. Growth factors such as transforming growth factor-β (TGF-β) are stored in a latent form within the myocardial interstitium and thereby form a reservoir of signaling molecules that directly influence myocardial ECM synthesis and degradation [14]. Moreover, mechanical stimuli such as stress or strain are likely transduced through the myocardial ECM to the cardiac myocyte, which in turn would directly affect myocyte growth. Thus, structural changes that would occur within the myocardial ECM would in turn affect myocyte biology and the overall structure and function of the myocardium.

Significant alterations in the structure and composition of the myocardial ECM occur following MI. Cardiac wound repair after MI involves temporarily overlapping phases, which include an inflammatory phase and tissue remodeling phase [14, 21]. The first phase starts after coronary artery occlusion with or without reperfusion and involves degradation of normal ECM, invasion of inflammatory cells at the site of initial injury, and the induction of bioactive peptides and cytokines.

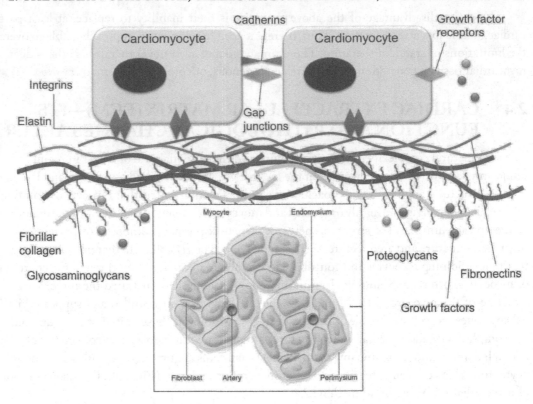

Figure 2.4: A schematic presentation of myocardial environment and major ECM components. Insert: Collagen weave surrounding individual myocytes and collagen struts tethering adjacent myocytes comprise the endomysium. Groups of myocytes are bundled within the perimysium. Capillaries and coronary microvessels have free diffusion access to cardiac myocytes throughout the ECM. In the scheme: major ECM components in the myocardium include fibrillar collagen, fibronectin, elastin, proteoglycans, and glycosaminoglycans. ECM also serves as controlled reservoir of growth factors. The interaction of ECM with cells is mediated by integrins on cell surface. Cadherins and gap junction proteins comprise the cell-cell interaction complex. Insert: reprinted with permission from [15].

Degradation of the ECM during the acute phase is considered to be an essential event that allows for the ingress of inflammatory cells as well as proliferation and maturation of macrophages and fibroblasts, and provides the necessary substructure for scar formation. Very early post-MI, there is a disappearance of the normal collagen matrix, increased release of hydroxyproline (an amino acid primarily found in collagen), and reduced collagen cross-linking within the ischemic region, all indicating that excessive ECM degradation takes place. These early ECM events occurred prior to the egress of inflammatory cells into the MI region. In this early time period, LV myocardial ECM

Table 2.1: The main components and function of myocardial extracellular matrix [16, 17]

Component	Main Function
Collagen fibrils (types I and III)	Structural support, maintain shape
	Transmission of force
	Tensile strength (type I); resilience (type III)
Elastin	Resilience; vessel wall stretch; cardiac wall stretch and relaxation
Hydrophilic glycosaminoglycans	Diffusion of nutrients, metabolites, growth factors, cytokines, etc.
Proteoglycans	Mechanics, fluid dynamics?
Integrins (matrix receptors)	Myocyte-fibroblast/ECM interactions, matrix remodeling
Fibronectin and laminin	Adhesive fibrous proteins
Cells	
Fibroblasts	Produce fibrillar collagen
	Convert to myofibroblasts after injury
Macrophages	Phagocytosis, inflammatory response
Other cells	Endothelial cells, smooth muscle cells, pericytes, neurons

degradation and remodeling were associated with an increased probability of rupture [22]. Thus, dynamic changes occur within the myocardial ECM in the initial and early phases of the post-MI period that directly affect the mechanical properties of the LV myocardium.

As the MI period progresses over the next several days, an influx of inflammatory cells into the injured myocardium occurs, which results in further proteolysis of cellular and ECM proteins. In addition, this inflammatory response causes proliferation and differentiation of fibroblasts and other interstitial cells, and the elaboration of bioactive molecules which contribute to a robust synthesis of ECM for the purposes of scar formation [14]. These changes within the MI region yield distinctive cellular and extracellular phenotypic changes. For example, the differentiation and proliferation of fibroblasts within the MI region demonstrate a unique protein signature and function to not only synthesize ECM proteins critical for scar formation, but also contribute to the biophysical properties of the scar itself and have been termed myofibroblasts. The later phase of post-MI remodeling results in ECM changes within all regions of the LV: the MI region, the viable myocardium within the border zone, and the remote region. Within the MI region, the newly formed ECM provides a means to tether viable myocyte fascicles and thereby forms a substrate to resist deformation from the intracavitary stresses generated during the cardiac cycle. Failure of ECM support has been associated with LV wall thinning and slippage of myocyte fascicles. This adverse remodeling process has been termed "infarct expansion" and occurs in the absence of additional myocyte injury or alterations in LV loading conditions. It has been postulated that an acceleration of ECM degradation occurs within the myocardium surrounding the MI (border zone) and facilitates the infarct expansion process in this later phase of post-MI remodeling [14]. This post-MI remodeling process is a clinically significant problem in that it can lead to LV dilation, systolic and diastolic dysfunction, and the progression to heart failure [15, 23, 24, 25]. Indeed, this process, which includes changes in ECM structure and composition, is an independent predictor of morbidity and mortality [26, 27].

2.5   CONGENITAL HEART DEFECTS

Birth defects are the number one cause of death in infants in the US, and heart-related defects are the most common fatal birth defects [28]. Structural myocardial defects are diagnosed in approximately 1% of all newborns, with a risk of sudden cardiac death 25 to 100 times that of young patients in the general population [29]. In many cases, major heart defects (discussed below) are associated with other types of birth defects, cardiac or non-cardiac.

Over the past 20 to 30 years, major advances have been made in the diagnosis and treatment of congenital heart disease in children. As a result, many children with this disease now survive to adulthood. In the US alone, the population of adults with congenital heart disease, either surgically corrected or uncorrected, is estimated to be increasing at a rate of about 5% per year [30].

Congenital heart disease is often divided into two types. The first type is cyanotic condition, where patients have arterial oxygen desaturation resulting from the shunting of systemic venous blood to the arterial circulation. The magnitude of shunting determines the severity of desaturation. Most children with cyanotic heart disease do not survive to adulthood without surgical intervention. Tetralogy of Fallot, the most common cyanotic congenital heart defect after infancy, is characterized by a large ventricular septal defect, an aorta that overrides the left and right ventricles, obstruction of the right ventricular outflow tract, and right ventricular hypertrophy. Most patients with tetralogy of Fallot have substantial right-to-left blood shunting and therefore have cyanosis (oxygen desaturation). Without surgical intervention, most patients die in childhood: the rate of survival is 66% at 1 year of age, and it declines to 11% at 20 years and 3% and 40 years [31].

The second type of congenital heart disease results from acyanotic conditions, where no significant oxygen desaturation occurs. The most common heart malformations here are ventricular or atrial septal defects. Ventricular septal defect is the most common congenital cardiac abnormality in infants and children. Ninety percent of the defects are eventually closed spontaneously by the time the child is 10 years old. As a result of the defect, there is a left-to-right blood shunting. Large defects usually lead to left ventricular failure or pulmonary hypertension with associated right ventricular failure [30]. Atrial septal defect accounts for about one third of the cases of congenital heart disease detected in adults. As a result of incomplete septal wall closure, the blood is shunted from one atrium to the other; the direction and magnitude of shunting are determined by the size of the defect and relative compliance of the ventricles. A sizable defect (more than 2 cm in diameter) may be associated with large shunt, with substantial hemodynamic consequences, that could lead to right ventricular dilation and failure over the years [30].

2.6 ENDOGENOUS MYOCARDIAL REGENERATION

For several decades, it has been assumed that the heart consists of only terminally differentiated cells and is incapable of intrinsically regenerating itself. However, recent data have challenged this view, and provide convincing evidence that new cardiomyocytes can be formed in the adult heart. Up-to-date, two main studies evaluated the degree of myocyte renewal in the adult human heart. Bergmann *et al* performed a virtual pulse-chase experiment by measuring the incorporation of carbon-14, which was released during an above-ground nuclear-bombs tests, into genomic DNA of human cardiomyocytes to calculate rates of turnover in these cells. The researchers found that at the age of 25 years, approximately 1% of cardiomyocytes turn over annually, and the turn-over rate decreases to 0.45% at the age of 75 years [32]. Kajstura *et al* utilized another approach, where they examined the incorporation of iododeoxyuridine in postmortem samples obtained from cancer patients who received this thymidine analog for therapeutic purposes. The authors report the average of 22% turnover of cardiomyocytes per year [33]. Despite the conflicting results, both reports point to the intrinsic regeneration ability of the adult human heart. Endogenous regeneration of the myocardium after the injury, such as MI or pressure overload, was reported in mice, where only 5-15% of remuscularization was detected [34]. The extremely low regeneration rates can explain the lack of significant restoration of cardiac muscle after severe ischemic injury, the formation of a fibrotic scar at the infarct, and progressive deterioration in cardiac function, that eventually lead to CHF.

Two major mechanisms could account for the endogenous regeneration of the myocardium [35]. By the first mechanism, adult cardiomyocytes may re-enter the cell cycle and divide. This type of myocyte regeneration represents an ancient regenerative program observed in the hearts of amphibians and fish. For instance, cardiomyocyte dedifferentiation and subsequent proliferation is the major mechanism of heart regeneration in zebrafish, rather than regeneration by progenitor or stem cells [36]. Importantly, cell cycle control in adult cardiomyocytes could be altered or reprogrammed in order to induce cell proliferation, either by genetic manipulation or by applying various bioactive molecules [37, 38, 39, 40, 41]. The second mechanism of endogenous regeneration is driven by resident populations of cardiac progenitor (CPC) or stem (CSC) cells. Numerous CSC/CPC pools within the adult heart have been characterized on the basis of stem cell marker expression and cardiomyogenic potential [42]. Nevertheless, despite the apparent existence of those subpopulations, the recruitment and/or activation of resident CSC/CPCs for cardiac repair is insufficient to significantly affect and prevent the deterioration in cardiac performance and adverse remodeling after a major ischemic event, due to the physical separation of CSC/CPC niches from the site of injury, the formation of fibrotic scar tissue, or the lack of appropriate signaling.

2.7   POTENTIAL THERAPEUTIC TARGETS AND STRATEGIES TO INDUCE MYOCARDIAL REGENERATION

In order to compensate for the low and insufficient intrinsic regeneration ability of the adult heart, the strategies for therapeutic regeneration aim to induce myocardial regeneration, improve tissue salvage, facilitate self-repair, reverse or attenuate adverse remodeling, and ultimately achieve long-term functional stabilization and improvement in heart function. Five major processes associated with MI are targeted at present by various experimental regeneration strategies [43]:

Cardioprotection – the prevention of progressive cardiomyocyte loss following MI by applying various apoptosis-inhibiting reagents or by inducing pro-survival signaling [44, 45].

Inflammation – time-adjusted modulation of the post-MI pro/anti-inflammatory cytokine/chemokine profile or cellular responses (e.g., granulation tissue formation and macrophage infiltration) in an attempt to induce effective tissue healing and repair and to avoid negative inflammatory effects (e.g., cell death, fibrosis, etc.) [46, 47, 48].

ECM remodeling and cardiac fibrosis – time-adjusted positive modulation of the fibrotic response (i.e., ECM remodeling and scar formation), utilizing recent knowledge on pro-fibrotic signaling, matrix metalloproteinase (MMP) inhibition, or modification of the MMP : tissue inhibitors of MMPs (TIMPs) ratio, which may lead to successful anti-fibrotic therapy [49, 50].

Angiogenesis – an effective tissue healing by increasing the blood supply to ischemic regions is an extensively used approach employed with a variety of strategies, proteins, genes, or cells, aimed at inducing the formation of new vasculature at the infarct site [51, 52].

Cardiomyogenesis – myocyte regeneration by activation and/or migration of distinct cell populations with stem- or progenitor-like properties in the adult myocardium which can contribute to *de novo* myocardium formation after MI. In addition, another novel mechanism for endogenous myocyte regeneration could be the induction of cardiomyocyte cell cycle re-entry by reprogramming of differentiated cardiomyocytes toward proliferation.

All these targets and goals can be translated into various therapy strategies aimed at inducing myocardial regeneration (Fig. 2.5).

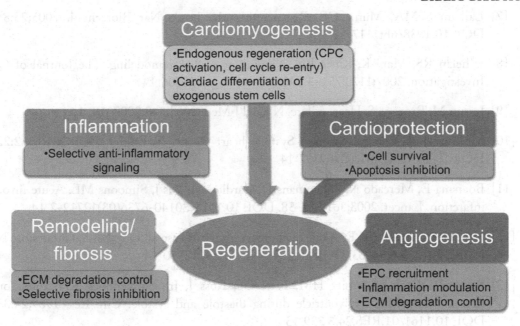

Figure 2.5: Targets and goals of therapeutic myocardial regeneration (see text for details).

BIBLIOGRAPHY

[1] Tortora GJ, Derrickson BH. Principles of Anatomy and Physiology, Twelfth Edition with Atlas and Registration Card Binder Ready Version: John Wiley & Sons; 2008. 8, 9, 10, 11

[2] Bers DM. Cardiac excitation-contraction coupling. Nature. 2002;415:198–205. DOI: 10.1038/415198a 10

[3] Gregorio CC, Antin PB. To the heart of myofibril assembly. Trends in cell biology. 2000;10:355–62. DOI: 10.1016/S0962-8924(00)01793-1 11

[4] World Health Organization - The Atlas of Heart Disease and Stroke. http://www.who.int/cardiovascular_diseases/resources/atlas/en/. 12

[5] Roger VL, Go AS, Lloyd-Jones DM, Benjamin EJ, Berry JD, Borden WB, et al. Heart Disease and Stroke Statistics–2012 Update: A Report From the American Heart Association. Circulation. 2012;125:e2-e220. DOI: 10.1161/CIR.0b013e31823ac046 12

[6] Cohen S, Leor J. Rebuilding broken hearts. Biologists and engineers working together in the fledgling field of tissue engineering are within reach of one of their greatest goals: constructing a living human heart patch. Scientific American. 2004;291:44–51. 13

[7] Laflamme MA, Murry CE. Regenerating the heart. Nat Biotechnol. 2005;23:845–56. DOI: 10.1038/nbt1117 12, 14

[8] Whelan RS, Mani K, Kitsis RN. Nipping at cardiac remodeling. The Journal of Clinical Investigation. 2007;117:2751–3. DOI: 10.1172/JCI33706 14

[9] Jessup M, Brozena S. Heart failure. N Engl J Med. 2003;348:2007–18. 14

[10] McMurray JJ. Clinical practice. Systolic heart failure. N Engl J Med. 2010;362:228–38. DOI: 10.1056/NEJMcp0909392 14

[11] Boersma E, Mercado N, Poldermans D, Gardien M, Vos J, Simoons ML. Acute myocardial infarction. Lancet. 2003;361:847–58. DOI: 10.1016/S0140-6736(03)12712-2 14

[12] Nabel EG, Braunwald E. A tale of coronary artery disease and myocardial infarction. N Engl J Med. 2012;366:54–63. DOI: 10.1056/NEJMra1112570 14

[13] Streeter DD, Jr., Spotnitz HM, Patel DP, Ross J, Jr., Sonnenblick EH. Fiber orientation in the canine left ventricle during diastole and systole. Circ Res. 1969;24:339–47. DOI: 10.1161/01.RES.24.3.339 15

[14] Spinale FG. Myocardial matrix remodeling and the matrix metalloproteinases: influence on cardiac form and function. Physiological reviews. 2007;87:1285–342. DOI: 10.1152/physrev.00012.2007 15, 17, 18

[15] Berk BC, Fujiwara K, Lehoux S. ECM remodeling in hypertensive heart disease. J Clin Invest. 2007;117:568–75. DOI: 10.1172/JCI31044 16, 18

[16] Jugdutt BI. Ventricular remodeling after infarction and the extracellular collagen matrix: when is enough enough? Circulation. 2003;108:1395–403. DOI: 10.1161/01.CIR.0000085658.98621.49 17

[17] Fomovsky GM, Thomopoulos S, Holmes JW. Contribution of extracellular matrix to the mechanical properties of the heart. J Mol Cell Cardiol. 2010;48:490–6. DOI: 10.1016/j.yjmcc.2009.08.003 17

[18] Ergul A, Walker CA, Goldberg A, Baicu SC, Hendrick JW, King MK, et al. ET-1 in the myocardial interstitium: relation to myocyte ECE activity and expression. American journal of physiology. 2000;278:H2050–6. 15

[19] Dell'Italia LJ, Meng QC, Balcells E, Wei CC, Palmer R, Hageman GR, et al. Compartmentalization of angiotensin II generation in the dog heart. Evidence for independent mechanisms in intravascular and interstitial spaces. J Clin Invest. 1997;100:253–8. DOI: 10.1172/JCI119529 15

[20] Flesch M, Hoper A, Dell'Italia L, Evans K, Bond R, Peshock R, et al. Activation and functional significance of the renin-angiotensin system in mice with cardiac restricted overexpression of tumor necrosis factor. Circulation. 2003;108:598–604. DOI: 10.1161/01.CIR.0000081768.13378.BF 15

[21] Dobaczewski M, Gonzalez-Quesada C, Frangogiannis NG. The extracellular matrix as a modulator of the inflammatory and reparative response following myocardial infarction. Journal of Molecular and Cellular Cardiology. 2010;48:504–11. DOI: 10.1016/j.yjmcc.2009.07.015 15

[22] Lerman RH, Apstein CS, Kagan HM, Osmers EL, Chichester CO, Vogel WM, et al. Myocardial healing and repair after experimental infarction in the rabbit. Circ Res. 1983;53:378–88. DOI: 10.1161/01.RES.53.3.378 17

[23] Iraqi W, Rossignol P, Angioi M, Fay R, Nuee J, Ketelslegers JM, et al. Extracellular cardiac matrix biomarkers in patients with acute myocardial infarction complicated by left ventricular dysfunction and heart failure: insights from the Eplerenone Post-Acute Myocardial Infarction Heart Failure Efficacy and Survival Study (EPHESUS) study. Circulation. 2009;119:2471–9. DOI: 10.1161/CIRCULATIONAHA.108.809194 18

[24] Kim HE, Dalal SS, Young E, Legato MJ, Weisfeldt ML, D'Armiento J. Disruption of the myocardial extracellular matrix leads to cardiac dysfunction. J Clin Invest. 2000;106:857–66. DOI: 10.1172/JCI8040 18

[25] Senzaki H, Paolocci N, Gluzband YA, Lindsey ML, Janicki JS, Crow MT, et al. beta-blockade prevents sustained metalloproteinase activation and diastolic stiffening induced by angiotensin II combined with evolving cardiac dysfunction. Circ Res. 2000;86:807–15. DOI: 10.1161/01.RES.86.7.807 18

[26] Chareonthaitawee P, Christian TF, Hirose K, Gibbons RJ, Rumberger JA. Relation of initial infarct size to extent of left ventricular remodeling in the year after acute myocardial infarction. J Am Coll Cardiol. 1995;25:567–73. DOI: 10.1016/0735-1097(94)00431-O 18

[27] White HD, Norris RM, Brown MA, Brandt PW, Whitlock RM, Wild CJ. Left ventricular end-systolic volume as the major determinant of survival after recovery from myocardial infarction. Circulation. 1987;76:44–51. DOI: 10.1161/01.CIR.76.1.44 18

[28] Kung HC, Hoyert DL, Xu J, Murphy SL. Deaths: final data for 2005. National vital statistics reports : from the Centers for Disease Control and Prevention, National Center for Health Statistics, National Vital Statistics System. 2008;56:1–120. 18

[29] Silka MJ, Hardy BG, Menashe VD, Morris CD. A population-based prospective evaluation of risk of sudden cardiac death after operation for common congenital heart defects. J Am Coll Cardiol. 1998;32:245–51. DOI: 10.1016/S0735-1097(98)00187-9 18

24 BIBLIOGRAPHY

[30] Brickner ME, Hillis LD, Lange RA. Congenital heart disease in adults. First of two parts. N Engl J Med. 2000;342:256–63. DOI: 10.1056/NEJM200002033420507 18, 19

[31] Brickner ME, Hillis LD, Lange RA. Congenital heart disease in adults. Second of two parts. N Engl J Med. 2000;342:334–42. DOI: 10.1056/NEJM200003303421324 18

[32] Bergmann O, Bhardwaj RD, Bernard S, Zdunek S, Barnabe-Heider F, Walsh S, et al. Evidence for cardiomyocyte renewal in humans. Science. 2009;324:98–102. DOI: 10.1126/science.1164680 19

[33] Kajstura J, Urbanek K, Perl S, Hosoda T, Zheng H, Ogorek B, et al. Cardiomyogenesis in the adult human heart. Circ Res. 2010;107:305–15. DOI: 10.1161/CIRCRESAHA.110.223024 19

[34] Hsieh PC, Segers VF, Davis ME, MacGillivray C, Gannon J, Molkentin JD, et al. Evidence from a genetic fate-mapping study that stem cells refresh adult mammalian cardiomyocytes after injury. Nature medicine. 2007;13:970–4. DOI: 10.1038/nm1618 19

[35] Parmacek MS, Epstein JA. Cardiomyocyte renewal. N Engl J Med. 2009;361:86–8. DOI: 10.1056/NEJMcibr0903347 19

[36] Jopling C, Sleep E, Raya M, Marti M, Raya A, Belmonte JC. Zebrafish heart regeneration occurs by cardiomyocyte dedifferentiation and proliferation. Nature. 2010;464:606–9. DOI: 10.1038/nature08899 19

[37] Novoyatleva T, Diehl F, van Amerongen MJ, Patra C, Ferrazzi F, Bellazzi R, et al. TWEAK is a positive regulator of cardiomyocyte proliferation. Cardiovasc Res. 2010;85:681–90. DOI: 10.1093/cvr/cvp360 19

[38] Bersell K, Arab S, Haring B, Kuhn B. Neuregulin1/ErbB4 signaling induces cardiomyocyte proliferation and repair of heart injury. Cell. 2009;138:257–70. DOI: 10.1016/j.cell.2009.04.060 19

[39] Hassink RJ, Pasumarthi KB, Nakajima H, Rubart M, Soonpaa MH, de la Riviere AB, et al. Cardiomyocyte cell cycle activation improves cardiac function after myocardial infarction. Cardiovasc Res. 2008;78:18–25. DOI: 10.1093/cvr/cvm101 19

[40] Campa VM, Gutierrez-Lanza R, Cerignoli F, Diaz-Trelles R, Nelson B, Tsuji T, et al. Notch activates cell cycle reentry and progression in quiescent cardiomyocytes. J Cell Biol. 2008;183:129–41. DOI: 10.1083/jcb.200806104 19

[41] Ahuja P, Sdek P, MacLellan WR. Cardiac myocyte cell cycle control in development, disease, and regeneration. Physiological reviews. 2007;87:521–44. DOI: 10.1152/physrev.00032.2006 19

[42] Bollini S, Smart N, Riley PR. Resident cardiac progenitor cells: At the heart of regeneration. J Mol Cell Cardiol. 2010;doi:10.1016/j.yjmcc.2010.07.006. DOI: 10.1016/j.yjmcc.2010.07.006 19

[43] Ruvinov E, Dvir T, Leor J, Cohen S. Myocardial repair: from salvage to tissue reconstruction. Expert Rev Cardiovasc Ther. 2008;6:669–86. DOI: 10.1586/14779072.6.5.669 20

[44] Abbate A, Bussani R, Amin MS, Vetrovec GW, Baldi A. Acute myocardial infarction and heart failure: role of apoptosis. Int J Biochem Cell Biol. 2006;38:1834–40. DOI: 10.1016/j.biocel.2006.04.010 20

[45] Garg S, Narula J, Chandrashekhar Y. Apoptosis and heart failure: clinical relevance and therapeutic target. J Mol Cell Cardiol. 2005;38:73–9. DOI: 10.1016/j.yjmcc.2004.11.006 20

[46] Nian M, Lee P, Khaper N, Liu P. Inflammatory cytokines and postmyocardial infarction remodeling. Circ Res. 2004;94:1543–53. DOI: 10.1161/01.RES.0000130526.20854.fa 20

[47] Frangogiannis NG, Smith CW, Entman ML. The inflammatory response in myocardial infarction. Cardiovasc Res. 2002;53:31–47. DOI: 10.1016/S0008-6363(01)00434-5 20

[48] Harel-Adar T, Ben Mordechai T, Amsalem Y, Feinberg MS, Leor J, Cohen S. Modulation of cardiac macrophages by phosphatidylserine-presenting liposomes improves infarct repair. Proceedings of the National Academy of Sciences of the United States of America. 2011;108:1827–32. DOI: 10.1073/pnas.1015623108 20

[49] Leask A. TGFbeta, cardiac fibroblasts, and the fibrotic response. Cardiovasc Res. 2007;74:207–12. DOI: 10.1016/j.cardiores.2006.07.012 20

[50] Vanhoutte D, Schellings M, Pinto Y, Heymans S. Relevance of matrix metalloproteinases and their inhibitors after myocardial infarction: a temporal and spatial window. Cardiovasc Res. 2006;69:604–13. DOI: 10.1016/j.cardiores.2005.10.002 20

[51] Renault MA, Losordo DW. Therapeutic myocardial angiogenesis. Microvasc Res. 2007;74:159–71. DOI: 10.1016/j.mvr.2007.08.005 20

[52] Tomanek RJ, Zheng W, Yue X. Growth factor activation in myocardial vascularization: therapeutic implications. Mol Cell Biochem. 2004;264:3–11. DOI: 10.1023/B:MCBI.0000044369.88528.a3 20

CHAPTER 3

Cell Sources for Cardiac Tissue Engineering

CHAPTER SUMMARY

Ideally, regeneration of the myocardial tissue after a major insult should involve adding new contracting cardiomyocytes into the infarct zone, which after integration with the host tissue would empower the heart contractility. Naturally, the best cell source for this purpose is fully differentiated and functional autologous cardiomyocytes. However, in a real world situation, the clinical applicability of this strategy is very limited since adult cardiomyocytes have lost their capability to proliferate and regenerate after damage. This chapter presents several clinically relevant cell sources that are in use or could potentially be used for cardiac tissue engineering and regeneration strategies. Cardiomyocytes derived from human embryonic stem cells, induced pluripotent stem cells or by direct reprogramming of somatic cells will be first introduced followed by the presentation of contemporary alternatives, such as autologous stem/progenitor cells purified from bone marrow, adipose tissue, or cardiac biopsies. The chapter concludes with a description of clinical studies performed with adult stem cells, their results, and the "paracrine theory" explaining the beneficial effects of stem/progenitor cell transplantation in improving cardiac function.

3.1   INTRODUCTION

In pathological situations where large numbers of cardiomyocytes are lost, e.g., following severe ischemic injury after MI or chronic stress, the endogenous regeneration capacity is insufficient to form adequate cardiac contractile mass to maintain heart contractility. Cardiac cell therapy, therefore, ideally aims at actively replacing the damaged and non-functional cardiomyocytes with a new and viable transplantable tissue. The characteristics of the ideal cell type have been articulated by many experts: a cell type should be both quantitatively and temporally available, safe to administer, effective at engraftment, and (most importantly) induce cardiac repair. Some argue that a source of autologous therapy is ideal so as to avoid any possibility of rejection, although it should be acknowledged that allogeneic cell therapy is also emerging as a strong possibility. Practical considerations, including the cost of therapy, will ultimately bear importantly on the accessibility of a new therapy [1].

Theoretically, the natural electro-physiological, structural, and contractile properties of differentiated cardiomyocytes make them the ideal candidate as cell source for *in vitro* bioengineering of cardiac tissue. Indeed, in most studies cardiac patches were produced using embryonic, fetal, or neonatal rat cardiomyocytes (see Chapter 6). However, cardiomyocytes are difficult to obtain and expand, are sensitive to ischemic insults, and are allogenic, that is, they can evoke immune response in the host tissue. The clinical need for human cardiomyocytes was at least partially fulfilled by introduction of human pluripotent stem cell-derived cardiomyocytes (either human embryonic or induced pluripotent stem cell-derived) (Fig. 3.1) [2, 3, 4, 5]. However, differentiation efficiency, immunogenicity, and various safety concerns (genome stability, mutations, and possible teratoma formation) are still major hurdles in the translation of applications of these cell types in clinics.

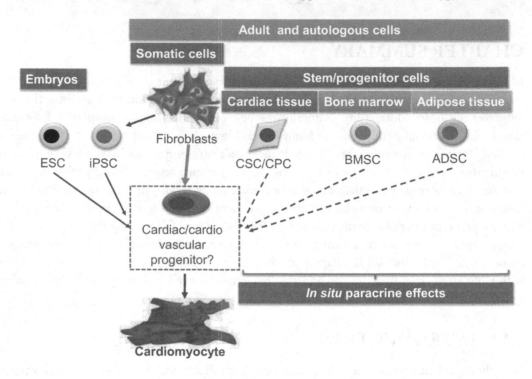

Figure 3.1: Clinically relevant cell sources for myocardial tissue engineering and regeneration. Blastocyst-derived embryonic stem cells (ESC) have an established potential to differentiate into cardiomyocytes. An autologous alternative to ESC is represented by introduction of induced pluripotent stem cells (iPSC) that can be derived from somatic cells (e.g., fibroblasts). Moreover, protocols are developed for direct reprogramming of somatic cells into cardiomyocytes. The differentiation process goes through several stages with varying efficiency, some leading to formation of cardiac or cardiovascular progenitors. Additional adult and autologous sources for stem/progenitor cells are the bone marrow, adipose, or cardiac tissues. The differentiation potential of these cells is contradictory, and mostly limited. Nevertheless, these cells were proven to exert beneficial paracrine effects in infarcted heart upon transplantation. CSC/CPC, cardiac stem/progenitor cells; BMSC, bone marrow-derived stem cells; ADSC, adipose tissue-derived stem cells. See text for more details.

Meanwhile, other adult stem or progenitor cell types with some indications of cardiomyogenic differentiation potential and/or with established beneficial positive effects on infarct repair represent a possible alternative, and in most cases, autologous source (i.e., bone-marrow-derived stem cell subsets, adipose-tissue derived, and cardiac progenitors) of cells that can be used for cardiac tissue engineering and myocardial regeneration after MI (Fig. 3.1) [1, 6, 7, 8, 9, 10, 11].

3.2   SOURCES FOR *DE NOVO* CARDIOMYOCYTES FOR CLINICAL APPLICATIONS

As mentioned, the vast majority of studies attempting to engineer cardiac patch *in vitro*, used rat neonatal, fetal, or embryonic cardiomyocytes as an ultimate cell source. Although these cells are easily accessible and represent a good model system for various *in vitro* studies, translation of cell transplantation, as well as graft engineering efforts into clinics, still suffers from the lack of appropriate cell source. As adult human cardiomyocytes are hard to obtain and have no ability to proliferate in culture, there is an urgent need to find an alternative suitable stem cell source, which will have a proven ability to effectively differentiate into functional cardiomyocytes. Not surprisingly, although having a promising potential for cardiogenesis, no cell type is perfect, each still having intrinsic drawbacks, as will be discussed below.

3.2.1   HUMAN EMBRYONIC STEM CELLS

Embryonic stem cells (ESCs) have emerged as one of the most promising sources of cardiac cells for transplantation purposes. Human ESCs (hESCs) first isolated in 1998 by James Thomson and co-workers from the inner cell mass of preimplantation embryos [12], are pluripotent cells capable of differentiating into virtually every cell type including cells of the heart [13, 14]. In the decade following the isolation of hESCs, protocols to differentiate these cells into cardiomyocytes have been refined [15]. Several groups have successfully isolated cardiomyocytes or cardiac progenitor cells from differentiating ESCs grown either in three-dimensional clumps termed embryoid bodies or 2D cultures treated with various extracellular proteins and/or growth factors and cytokines (i.e., BMP-4, activin A) that increase the yield of cardiac cells. Importantly, the ESC-derived cardiomyocytes not only share molecular markers with primary cardiomyocytes, but ultrastructural (electron microscopy), electrophysiological (action potential measurements), and mechanical (determination of contractility) studies of the ESC progeny indicate that they also exhibit all hallmarks of cardiomyocytes. Of potential importance, ESC-derived cardiomyocytes have been shown to exhibit a phenotype reminiscent of fetal, rather than adult cardiomyocytes. Given the observed differences between fetal and adult cardiomyocytes, ESC-derived cardiomyocytes with features of adult cells would probably be preferable for clinical transplantation purposes. Nonetheless, ESC-derived cardiomyocytes have already been used in transplantation experiments in rodent models of cardiac diseases [14]. These results show that hESC-derived cardiomyocytes can couple electromechanically with cardiomyocytes of the host, and therapeutic effects in the MI model have been reported four weeks after transplantation [3]. However, in a study with a longer follow up, no effect on cardiac function could be documented twelve weeks after transplantation [16]. Thus, the long-term effects of ESC-derived cardiomyocytes to injured myocardium need to be evaluated further.

Two of the obstacles that stand in the way of the therapeutic use of ESC are immunological rejection (due to non-autologous nature) and the propensity of undifferentiated ESCs to form teratomas when injected *in vivo* [10]. As knowledge of pathways for ESC differentiation increases, cell differentiation will become more controllable, and this, together with effective selection protocols, could limit teratoma formation.

3.2.2   INDUCED PLURIPOTENT STEM CELLS

Recently, a novel way of generating stem cells from differentiated cells has been described. This technique, pioneered by Shinya Yamanaka and colleagues, relies on the reprogramming of fully differentiated somatic cells to ESC-like cells, known as induced pluripotent stem cells (iPSCs) [5, 17]. This conversion was achieved by lentiviral-based transduction of four transcriptional factors, Oct3/4, Sox2, Klf4, and c-Myc. IPSCs exhibit the two key features of ESCs, in that they can be expanded over many passages *in vitro* and give rise to cells of all three germ layers, both under appropriate *in vivo* and *in vitro* differentiation conditions. Originally established for mouse embryonic fibroblasts with a genetic selection strategy for identifying reprogrammed cells, the basic iPSC derivation protocol has been refined by many groups, permitting reprogramming of human cells, reprogramming without genetic selection and chemicals enhancing reprogramming efficiency [13, 15]. The ability of both mouse and human iPSC to differentiate into functional cardiomyocytes has recently been demonstrated [4, 18, 19, 20, 21]. IPSCs were shown to differentiate efficiently into cardiomyocytes with cardiac-specific molecular, structural, and functional properties that recapitulate the developmental ontogeny of cardiogenesis. The study of iPSC from selected cohorts of patients was found to be a very efficient way to uncover molecular mechanisms of disease. This was shown, for instance, by iPSC generation from patients with long QT syndrome, with subsequent differentiation of iPSC to cardiomyocytes, which recapitulated the electrophysiological features of the disorder [22, 23].

Similar to ESC-derived cardiomyocytes, iPSC-derived cells are largely immature and most analogous to fetal stages of development, exhibiting automaticity (spontaneous contraction), fetal-type ion channel expression, fetal-type gene expression patterns, and fetal-type physical phenotypes [15]. Three major subtypes of ESC or iPSC-derived cardiomyocytes can be derived that have atrial-, ventricular-, or nodal-like phenotypes as determined by electrophysiological analysis of action potentials, and the specific required type can be potentially enriched [15].

Being derived from adult cells, iPSCs bypass the ethical issues regarding the use of embryonic human tissue to cure disease, and immunocompatibility is not an issue because the starting material, i.e., skin fibroblasts, can be obtained from the patient. However, there are currently caveats with the iPSC reprogramming procedure that need to be addressed before this elegant technology can be put to clinical use. One important aspect is that the original protocol for reprogramming of human cells to iPSCs relies on the use of viruses integrating into the genome of cells undergoing the reprogramming process. Clinical trials in gene therapy have shown that integration of viruses in tumor-suppressor genes may give a selective advantage and thereby promote malignancy when transplanted to the patient. Moreover, given that some of the virally encoded genes are oncogenes that may be reactivated after transplantation, it is clear that protocols permitting reprogramming without the use of viruses are essential before iPSCs can become a clinical tool [13, 15]. Furthermore, the long-term performance of differentiated cells derived from iPSCs needs to be critically assessed. In this regard, a detailed comparison between cardiomyocytes derived from iPSCs and cardiomyocytes derived from ESCs is required before iPSCs can be considered for regenerative therapy in cardiology. Even with an efficient method for nonviral iPSC generation of human cells in place, and even if iPSCs prove to be functionally equivalent to human ESCs, there is still one potentially important drawback with the iPSC technology–the time for iPSC derivation, with stringent criteria for pluripotency and capacity to differentiate to the lineage of interest, requires time which would at least be relevant for situations in which patient-specific lines are sought. This requirement is in sharp contrast to banked hESCs, which would be relatively easy to expand, store, and distribute in an off-the-shelf paradigm. For use of iPSCs in a similar manner, banks of pretested patient-specific pluripotent cell lines would have to be established [13].

3.2.3   DIRECT REPROGRAMMING OF DIFFERENTIATED SOMATIC CELLS

Although still in its infancy, it appears likely that cellular reprogramming may provide important tools for translational scientists aiming at generating a specific cell type for cell therapy. In addition to the iPSC technology, where fully differentiated cells are reprogrammed to the fully undifferentiated ESC-like state, and subsequent differentiation of such cells can give the desired cell type, one can envision more direct ways of reprogramming cells. In a pioneering experiment conducted more than 20 years ago, Harold Weintraub and colleagues showed that forced expression of the myogenic transcription factor *MyoD* in cultured fibroblasts caused such cells to adopt the myocyte fate [24]. This process is comparatively simple because *MyoD* acts as a master regulator of skeletal muscle formation, whereas it would appear that no such single gene exists for the direct reprogramming of cardiomyocytes [15]. With years, work on reprogramming to pluripotency has provided a new insight into direct reprogramming, combining high-expression retroviral vectors with the subtractive assessment strategy. Implementing these advances, Ieda and colleagues successfully reprogrammed cardiac fibroblasts into cardiomyocyte-like cells (referred to as iCMs) via the exogenous expression of *Gata4*, *Mef2c*, and *Tbx5* (GMT cocktail) [25]. The researchers showed that this reprogramming was not achieved via pluripotent intermediate, but was the result of gradual loss of fibroblast identity and progressive upregulation of cardiomyocyte-specific genes. ICMs were induced *in vitro* within three days, although the reprogramming factors were shown to be required for two weeks for stable reprogramming. The iCMs produced electrophysiological and gene expression profiles similar to those of fetal cardiomyocytes, although only 30% of the iCMs (approximately 6% of the total cell number) exhibited spontaneous contraction. The GMT cocktail was also shown to be able to reprogram adult tail-tip dermal fibroblasts to iCMs to a similar efficiency, albeit with lower expression of cardiac Troponin T and no demonstration of spontaneous contraction or electrophysiological properties. Transplantation of neonatal cardiac fibroblasts into mouse hearts one day after reprogramming showed that reprogramming can be achieved *in vivo* [25]. More recently, the same group reported direct reprogramming of resident non-myocytes in the murine heart into cardiomyocyte-like cells (shown to have assembled sarcomeres, cardiomyocyte-like gene expression, ventricular-like action potentials, and electrical coupling) *in vivo* by local delivery of GMT after coronary ligation. This strategy also resulted in decreased infarct size and modestly attenuated cardiac dysfunction up to three months after MI induction [26]. For a similar goal, Jayawardena *et al* used another emerging approach, a combination of three specific micro (mi)RNAs(-1, -133, and -208), to induce direct reprogramming of murine fibroblasts to cardiomyocyte-like cells *in vitro* and *in vivo* [27].

The main drawbacks of direct reprogramming are the low efficiency rates (1 in 20 fibroblasts are successfully reprogrammed) at present, the use of randomly integrating retroviruses (which preclude future clinical application), the experimental complexity (which will make the GMP production process challenging), and the potential for contamination by endogenous cardiomyocytes; most importantly, use of this methodology has not yet been independently verified or demonstrated in human cells to date. The relationship between the properties of iCMs and genuine cardiomyocytes has also not been fully established using more stringent criteria. Finally, a definitive proof of this methodology lies in reprogramming skin fibroblasts (and various other cell types) into cardiac cells, as can be done reproducibly for the reprogramming of various adult somatic cell types to pluripotency [15].

3.3 CONTEMPORARY ALTERNATIVES—ADULT AUTOLOGOUS STEM AND PROGENITOR CELLS

3.3.1 BONE MARROW-DERIVED STEM CELLS

The bone marrow in adults is a complex organ that harbors numerous types of mature and immature hematopoietic and non-hematopoietic cells. Consistent with the notion that various adult organs harbor tissue-specific progenitors that give rise to cells with adult phenotypes continuously or following organ damage, stem/progenitor cells with the potential to repair diverse tissues have been well described in the bone marrow. Various BM cells have been examined for their ability to repair myocardial damage, and initial animal studies to date have shown therapeutic benefits.

First population is the bone marrow mononuclear cells (BMMNCs) that represent a heterogeneous cell population that contains hematopoietic and non-hematopoietic cells with diverse phenotypes. BMMNCs are generally isolated from total BMCs by density gradient centrifugation, which allows separation of BMMNCs relatively easily and quickly. BMMNCs are relatively easy to procure in large numbers and do not require complex culture conditions [28].

Another widely used cell population residing in the bone marrow is mesenchymal stem cells (MSC) that are the multipotent precursors of various non-hematopoietic lineages and possess the ability to differentiate into adipose, bone, cartilage, skeletal muscle, neural, and other phenotypes.

Several studies have also documented the ability of MSCs to differentiate into cardiomyocytes *in vitro* and following transplantation into the infarcted myocardium *in vivo*. However, the cardiomyogenic potential of MSCs and other bone-marrow derived stem cells is still highly controversial [29]. Nevertheless, MSCs still remain an attractive potential cell source, also suitable for preemptive harvest, rapid expansion *in vitro*, and prolonged storage for future use, perhaps as an off-the-shelf product [28].

3.3.2 ADIPOSE TISSUE-DERIVED STEM CELLS

Fat is abundant in most individuals, allowing a simpler and more efficient harvesting, as adipose tissue has a higher stem cell yield than bone marrow, and diminishing the need of *in vitro* expansion [30]. Adipose-derived stem cells (ADSCs) can easily be isolated and cultured *ex vivo* and express markers associated with mesenchymal and perivascular cells, maintaining their characteristic multipotency to differentiate into chondrocytes, osteoblasts, endothelial cells, and cardiomyocytes. The differentiation capacity and paracrine activity of these cells made them an optimal candidate for the treatment of a diverse range of diseases: from immunological disorders, such as graft versus host disease, to cardiovascular pathologies, such as peripheral ischemia. Four different possible fates of ADSCs are described: (i) differentiating (rare and for a very limited degree) into cardiac muscles by direct contact with adjacent CM; (ii) differentiating into SMCs that have migrated to and surrounded immature vessels; (iii) adipogenic differentiation; and (iv) secreting proangiogenic and other beneficial factors to recruit endogenous endothelial cells and positively affect infarct repair [31, 32].

3.3.3 CARDIAC STEM/PROGENITOR CELLS

Another possible source of cells for cardiac engineering is resident cardiac stem/progenitor cells. As mentioned, over the past decade, compelling evidence has arisen suggesting the heart has intrinsic regenerative potential [33, 34]. One of the mechanisms of endogenous myocardial regeneration could be driven to some extent by resident populations of cardiac progenitor cells (CPCs). Numerous CPC pools within the adult heart have been characterized on the basis of stem cell marker expression and cardiomyogenic potential [35]. These include: side-population stem cells, c-kit$^+$, Sca-1$^+$, Islet-1$^+$, SSEA-1$^+$, "cardiospheres" (CS), and/or cardiosphere-derived cells (CDC).

In humans, endogenous c-kit$^+$ CPCs have been described in patients undergoing gender-mismatched heart transplantation and in patients with aortic stenosis. Cardiomyogenic c-kit$^+$, Sca-1$^+$, and MDR1$^+$ progenitors have been found in the human heart, mainly in the atria, showing a high level of proliferation, marked commitment to the myocyte lineage, and positive expression of cardiovascular markers such as MEF2, Gata4, Nestin, and Flk-1, but negative for the expression of hematopoietic antigens. Human c-kit$^+$ CPCs were isolated from myocardial biopsies by enzymatic digestion and immuno-magnetic sorting. These cells demonstrated stem cell properties of self-renewal, clonogenicity, and the ability to give rise to all three cardiovascular lineages [36]. Other sources of human CPCs, relevant for translational medicine, are the human CS and CDCs. CS were first isolated as progenitors that spontaneously arose from cultures of postnatal murine heart explants and ventricular human biopsies. These highly clonogenic cells expressed stem and endothelial progenitor markers, such as c-kit, Sca-1, CD31, and CD34, and appeared to comprise a mixture of cardiovascular stem cells at different stages of commitment, including differentiating progenitors, spontaneously differentiated cardiomyocytes, vascular cells, and supporting cells of mesenchymal origin [35]. CS could also be cultured as a monolayer of cardiosphere-derived cells (CDCs), upon mechanical dissociation, and expanded over many passages. Smith and co-workers described a method to efficiently isolate human CDCs (described as a mixture of cardiac stem cells and supporting cells) from endomyocardial biopsy specimens, and provided proof of their highly cardiogenic properties *in vitro* and *in vivo* [37].

In a recent study, Li *et al* performed a head-to-head comparison of different stem cell types/subpopulations for functional myocardial repair by assessing multiple *in vitro* parameters, including secretion of relevant growth factors, and *in vivo* cell implantation into acute MI model in severe combined immunodeficiency (SCID) mice [38]. Human CDCs, BM-derived MSCs, AD-SCs, and BMMNCs were compared. Characterization of cell phenotypes revealed a distinctive phenotype in CDCs with uniform expression of CD105, partial expression of c-kit and CD90, and negligible expression of hematopoietic markers. On the basis of these findings, CDCs seem to contain a minority of fibroblast and/or weakly committed hematopoietic cells, whereas such populations dominate in the cells of bone marrow and adipose origins. *In vitro*, CDCs showed the greatest spontaneous myogenic differentiation potency (shown by the highest expression of Troponin-T), highest angiogenic potential (in *in vitro* tube-forming assay), and relatively high production of various angiogenic and antiapoptotic-secreted factors (angiopoietin-2, bFGF, HGF, IGF-1, SDF-1, and VEGF). *In vivo*, injection of CDCs into the infarcted mouse hearts resulted in superior improvement of cardiac function (in terms of increased left ventricle ejection fraction (LV EF)), the highest cell engraftment and myogenic differentiation rates, and the least-abnormal heart morphology three weeks after treatment. CDC-treated hearts also exhibited the lowest number of apoptotic cells. Interestingly, purified c-kit-positive fraction of CDCs was found less potent than unselected CDC mixture, suggesting that the presence of mesenchymal and stromal cell populations may favor endogenous regeneration and enhance overall paracrine potency [38]. In conclusion, CDCs exhibited a balanced profile of paracrine factor production and, among various comparator cell types/subpopulations, provided the greatest functional benefit in experimental MI.

3.4   CLINICAL TRIALS AND "PARACRINE EFFECT" HYPOTHESIS OF STEM/PROGENITOR CELL TRANSPLANTATION

The success in preclinical animal studies, in terms of infarct reduction, angiogenesis, and functional improvement, has powered the fast translation of stem cell transplantation to human trials. However, the translational application of candidate cell types has been criticized by some as being applied too rapidly before the cell biology is fully understood [39].

Despite the big promise seen in stem cell therapy for acute MI, the randomized controlled clinical trials in patients, beyond showing an apparent safety of the treatment, have resulted in only modest improvements, and some relatively large trials even failed to show any functional benefit (e.g., LV ejection fraction (EF) increase) [11, 40]. Although the several trials performed differ greatly in cell preparation protocols, timing of the treatment, routes of delivery, and patient characteristics, several major conclusions could be drawn based on the observed results. First, the efficacy of stem cell therapy is suboptimal due to extremely low engraftment rates of the transplanted cells [10]. In addition, transient improvements in cardiac function cannot be treated as direct evidence of cardiac regeneration *per se*, and the transplanted cells have no true cardiomyogenic differentiation potential [41]. Moreover, a portion of the positive effect of cell transplantation may relate to effects of decreasing wall stress by increasing tissue mass in a thinning myocardial wall, an anatomic and mechanistic effect that is independent of real regenerative effect [8, 9, 13].

In an attempt to achieve actual myocardial regeneration, two randomized phase 1 trials examined the feasibility of intracoronary delivery of autologous cardiac progenitor cell populations: SCIPIO study, using c-kit-positive cardiac stem cells, and CADUCEUS study using cardiosphere-derived cells [42, 43]. The SCIPIO study provided data showing improvement in left ventricular function with a 24% relative infarct size reduction and an 8.2% absolute improvement in EF after 4 months, in stem cell-treated patients. In contrast, the CADUCEUS study failed to show significant function improvement, but, nevertheless, detailed assessment of cardiac structure by magnetic resonance imaging (MRI) provided striking evidence of cardiac regeneration after cardiosphere-derived cell therapy, showing a reduction of 28% in average scar mass at 6 months and 42% at 12 months. Although direct comparisons cannot be made between the relative therapeutic efficacy of c-kit-positive cells and cardiosphere-derived cells because of the differences in study design and patient populations, the magnitude of reduction in relative infarct size shown in the two trials is not dissimilar and seems to improve with increased follow-up [44]. As proposed by the authors of the CADUCEUS study, the benefit of cardiosphere-derived cell therapy is probably mediated through activation of endogenous reparative and regenerative pathways or, to a lesser extent, through direct transdifferentiation of transplanted cardiosphere-derived cells into functional myocardium. Therefore, the efficacy of different populations of cardiac progenitor cells might be because such cells activate the same or similar endogenous mechanisms. It should be noted, however, that the observed results should be taken with caution, as both studies presented a small number of patients, some taken from non-randomized stages of the trial, and the trials are still ongoing.

The mixed results of stem cell transplantation trials in humans enforce rethinking and a more detailed analysis of the positive effects and mechanisms of action of cell transplantation. In this context, emerging evidence suggests that paracrine effects significantly contribute to the positive effect of cell therapy. The so-called "paracrine effect hypothesis" has already been confirmed in various stem cell types, and also deduced from the analysis of data from preclinical and clinical studies. These data show that the expression and secretion of various soluble factors from transplanted cells (cytokines, chemokines, growth factors, and others) could be responsible for the major mechanisms involved in myocardial repair, such as cell survival, improved contractility, neovascularization, differentiation and/or induction of endogenous regeneration, and more favorable remodeling [11, 45, 46, 47]. Although the identification of such a "cocktail" of components secreted from the transplanted cells is difficult, attempts have already been made to use known cardioprotective molecules to influence these mechanisms. Local intramyocardial delivery of such molecules could be more reproducible, less time consuming, and more technically appealing than the injection of heterogeneous populations of stem or progenitor cells (see Chapter 10).

3.5 SUMMARY AND CONCLUSIONS

Long-term functional improvement after MI depends on the restoration and regeneration of new myocardial tissue. Ideally, the best cell source for this task should be fully matured human cardiomyocytes; however, their generation represents a major challenge. This chapter presented strategies developed to promote the differentiation of pluripotent stem cells (ESC and iPSC) toward cardiomyocytes. The recent introduction of direct reprogramming has expanded the possibilities for generation of human cardiomyocytes from autologous somatic cells. Yet, the translation of these cells into clinical applications depends on solving several difficulties, such as the low conversion efficiency, and possible immunogenicity and tumorogenicity of the cells. Adult stem/progenitor cells represent a logical contemporary alternative. Although lacking the robust potential to differentiate into cardiomyocytes, these cell types have several advantages, such as relatively easy isolation and expansion techniques, as well as showing safety in clinical trials. Importantly, these cell types have shown beneficial, and mainly paracrine effects, in myocardial repair in animals. The translation to clinics, however, has been less effective, emphasizing the need for more careful evaluation of cellular effects on one hand, and promoting the use of sole paracrine factors, such as growth factors and cytokines, on the other.

BIBLIOGRAPHY

[1] Heldman AW, Zambrano JP, Hare JM. Cell therapy for heart disease: where are we in 2011? J Am Coll Cardiol. 2011;57:466–8. DOI: 10.1016/j.jacc.2010.09.028 27, 29

[2] Kehat I, Kenyagin-Karsenti D, Snir M, Segev H, Amit M, Gepstein A, et al. Human embryonic stem cells can differentiate into myocytes with structural and functional properties of cardiomyocytes. J Clin Invest. 2001;108:407–14. DOI: 10.1172/JCI12131 28

[3] Laflamme MA, Chen KY, Naumova AV, Muskheli V, Fugate JA, Dupras SK, et al. Cardiomyocytes derived from human embryonic stem cells in pro-survival factors enhance function of infarcted rat hearts. Nat Biotechnol. 2007;25:1015–24. DOI: 10.1038/nbt1327 28, 30

[4] Burridge PW, Thompson S, Millrod MA, Weinberg S, Yuan X, Peters A, et al. A universal system for highly efficient cardiac differentiation of human induced pluripotent stem cells that

eliminates interline variability. PLoS One. 2011;6:e18293. DOI: 10.1371/journal.pone.0018293 28, 30

[5] Takahashi K, Tanabe K, Ohnuki M, Narita M, Ichisaka T, Tomoda K, et al. Induction of pluripotent stem cells from adult human fibroblasts by defined factors. Cell. 2007;131:861–72. DOI: 10.1016/j.cell.2007.11.019 28, 30

[6] Nunes SS, Song H, Chiang CK, Radisic M. Stem Cell-Based Cardiac Tissue Engineering. J Cardiovasc Transl Res. 2011. DOI: 10.1007/s12265-011-9307-x 29

[7] Martinez EC, Kofidis T. Adult stem cells for cardiac tissue engineering. J Mol Cell Cardiol. 2011;50:312–9. DOI: 10.1016/j.yjmcc.2010.08.009 29

[8] Chavakis E, Koyanagi M, Dimmeler S. Enhancing the outcome of cell therapy for cardiac repair: progress from bench to bedside and back. Circulation. 2010;121:325–35. DOI: 10.1161/CIRCULATIONAHA.109.901405 29, 35

[9] Dimmeler S, Burchfield J, Zeiher AM. Cell-based therapy of myocardial infarction. Arterioscler Thromb Vasc Biol. 2008;28:208–16. DOI: 10.1161/ATVBAHA.107.155317 29, 35

[10] Segers VF, Lee RT. Stem-cell therapy for cardiac disease. Nature. 2008;451:937–42. DOI: 10.1038/nature06800 29, 30, 35

[11] Tongers J, Losordo DW, Landmesser U. Stem and progenitor cell-based therapy in ischaemic heart disease: promise, uncertainties, and challenges. Eur Heart J. 2011;32:1197–206. DOI: 10.1093/eurheartj/ehr018 29, 35, 36

[12] Thomson JA, Itskovitz-Eldor J, Shapiro SS, Waknitz MA, Swiergiel JJ, Marshall VS, et al. Embryonic stem cell lines derived from human blastocysts. Science. 1998;282:1145–7. DOI: 10.1126/science.282.5391.1145 29

[13] Hansson EM, Lindsay ME, Chien KR. Regeneration next: toward heart stem cell therapeutics. Cell Stem Cell. 2009;5:364–77. DOI: 10.1016/j.stem.2009.09.004 29, 30, 31, 35

[14] Passier R, van Laake LW, Mummery CL. Stem-cell-based therapy and lessons from the heart. Nature. 2008;453:322–9. DOI: 10.1038/nature07040 29

[15] Burridge PW, Keller G, Gold JD, Wu JC. Production of de novo cardiomyocytes: human pluripotent stem cell differentiation and direct reprogramming. Cell Stem Cell. 2012;10:16–28. DOI: 10.1016/j.stem.2011.12.013 29, 30, 31, 32

[16] van Laake LW, Passier R, Monshouwer-Kloots J, Verkleij AJ, Lips DJ, Freund C, et al. Human embryonic stem cell-derived cardiomyocytes survive and mature in the mouse heart and transiently improve function after myocardial infarction. Stem cell research. 2007;1:9–24. DOI: 10.1016/j.scr.2007.06.001 30

[17] Takahashi K, Yamanaka S. Induction of pluripotent stem cells from mouse embryonic and adult fibroblast cultures by defined factors. Cell. 2006;126:663–76. DOI: 10.1016/j.cell.2006.07.024 30

[18] Mauritz C, Schwanke K, Reppel M, Neef S, Katsirntaki K, Maier LS, et al. Generation of functional murine cardiac myocytes from induced pluripotent stem cells. Circulation. 2008;118:507–17. DOI: 10.1161/CIRCULATIONAHA.108.778795 30

[19] Zhang J, Wilson GF, Soerens AG, Koonce CH, Yu J, Palecek SP, et al. Functional cardiomyocytes derived from human induced pluripotent stem cells. Circ Res. 2009;104:e30–41. DOI: 10.1161/CIRCRESAHA.108.192237 30

[20] Mehta A, Chung YY, Ng A, Iskandar F, Atan S, Wei H, et al. Pharmacological response of human cardiomyocytes derived from virus-free induced pluripotent stem cells. Cardiovasc Res. 2011;91:577–86. DOI: 10.1093/cvr/cvr132 30

[21] Fujiwara M, Yan P, Otsuji TG, Narazaki G, Uosaki H, Fukushima H, et al. Induction and enhancement of cardiac cell differentiation from mouse and human induced pluripotent stem cells with cyclosporin-A. PLoS One. 2011;6:e16734. DOI: 10.1371/journal.pone.0016734 30

[22] Moretti A, Bellin M, Welling A, Jung CB, Lam JT, Bott-Flugel L, et al. Patient-specific induced pluripotent stem-cell models for long-QT syndrome. N Engl J Med. 2010;363:1397–409. DOI: 10.1056/NEJMoa0908679 30

[23] Itzhaki I, Maizels L, Huber I, Zwi-Dantsis L, Caspi O, Winterstern A, et al. Modelling the long QT syndrome with induced pluripotent stem cells. Nature. 2011;471:225–9. DOI: 10.1038/nature09747 30

[24] Davis RL, Weintraub H, Lassar AB. Expression of a single transfected cDNA converts fibroblasts to myoblasts. Cell. 1987;51:987–1000. DOI: 10.1016/0092-8674(87)90585-X 31

[25] Ieda M, Fu JD, Delgado-Olguin P, Vedantham V, Hayashi Y, Bruneau BG, et al. Direct reprogramming of fibroblasts into functional cardiomyocytes by defined factors. Cell. 2010;142:375–86. DOI: 10.1016/j.cell.2010.07.002 31, 32

[26] Qian L, Huang Y, Spencer CI, Foley A, Vedantham V, Liu L, et al. In vivo reprogramming of murine cardiac fibroblasts into induced cardiomyocytes. Nature. 2012. DOI: 10.1038/nature11044 32

[27] Jayawardena TM, Egemnazarov B, Finch EA, Zhang L, Payne JA, Pandya K, et al. MicroRNA-Mediated In Vitro and In Vivo Direct Reprogramming of Cardiac Fibroblasts to Cardiomyocytes. Circ Res. 2012. DOI: 10.1161/CIRCRESAHA.112.269035 32

[28] Dawn B, Abdel-Latif A, Sanganalmath SK, Flaherty MP, Zuba-Surma EK. Cardiac repair with adult bone marrow-derived cells: the clinical evidence. Antioxidants & redox signaling. 2009;11:1865–82. DOI: 10.1089/ars.2009.2462 32, 33

[29] Dawn B, Bolli R. Adult bone marrow-derived cells: regenerative potential, plasticity, and tissue commitment. Basic Res Cardiol. 2005;100:494–503. DOI: 10.1007/s00395-005-0552-5 33

[30] Mazo M, Gavira JJ, Pelacho B, Prosper F. Adipose-derived stem cells for myocardial infarction. J Cardiovasc Transl Res. 2011;4:145–53. DOI: 10.1007/s12265-010-9246-y 33

[31] Choi YS, Matsuda K, Dusting GJ, Morrison WA, Dilley RJ. Engineering cardiac tissue in vivo from human adipose-derived stem cells. Biomaterials. 2010;31:2236–42. DOI: 10.1016/j.biomaterials.2009.11.097 33

[32] Venugopal JR, Prabhakaran MP, Mukherjee S, Ravichandran R, Dan K, Ramakrishna S. Biomaterial strategies for alleviation of myocardial infarction. J R Soc Interface. 2012;9:1–19. DOI: 10.1098/rsif.2011.0301 33

[33] Bergmann O, Bhardwaj RD, Bernard S, Zdunek S, Barnabe-Heider F, Walsh S, et al. Evidence for cardiomyocyte renewal in humans. Science. 2009;324:98–102. DOI: 10.1126/science.1164680 33

[34] Kajstura J, Urbanek K, Perl S, Hosoda T, Zheng H, Ogorek B, et al. Cardiomyogenesis in the adult human heart. Circ Res. 2010;107:305–15. DOI: 10.1161/CIRCRESAHA.110.223024 33

[35] Bollini S, Smart N, Riley PR. Resident cardiac progenitor cells: At the heart of regeneration. J Mol Cell Cardiol. 2010;doi:10.1016/j.yjmcc.2010.07.006. DOI: 10.1016/j.yjmcc.2010.07.006 33, 34

[36] Bearzi C, Rota M, Hosoda T, Tillmanns J, Nascimbene A, De Angelis A, et al. Human cardiac stem cells. Proceedings of the National Academy of Sciences of the United States of America. 2007;104:14068–73. DOI: 10.1073/pnas.0706760104 33

[37] Smith RR, Barile L, Cho HC, Leppo MK, Hare JM, Messina E, et al. Regenerative potential of cardiosphere-derived cells expanded from percutaneous endomyocardial biopsy specimens. Circulation. 2007;115:896–908. DOI: 10.1161/CIRCULATIONAHA.106.655209 34

[38] Li TS, Cheng K, Malliaras K, Smith RR, Zhang Y, Sun B, et al. Direct comparison of different stem cell types and subpopulations reveals superior paracrine potency and myocardial repair efficacy with cardiosphere-derived cells. J Am Coll Cardiol. 2012;59:942–53. DOI: 10.1016/j.jacc.2011.11.029 34

[39] Chien KR. Stem cells: lost in translation. Nature. 2004;428:607–8. DOI: 10.1038/nature02500 34

[40] Menasche P. Cardiac cell therapy: Lessons from clinical trials. J Mol Cell Cardiol. 2010;doi:10.1016/j.yjmcc.2010.06.010. DOI: 10.1016/j.yjmcc.2010.06.010 35

[41] Laflamme MA, Murry CE. Heart regeneration. Nature. 2011;473:326–35. DOI: 10.1038/nature10147 35

[42] Bolli R, Chugh AR, D'Amario D, Loughran JH, Stoddard MF, Ikram S, et al. Cardiac stem cells in patients with ischaemic cardiomyopathy (SCIPIO): initial results of a randomised phase 1 trial. Lancet. 2011;378:1847–57. DOI: 10.1016/S0140-6736(11)61590-0 35

[43] Makkar RR, Smith RR, Cheng K, Malliaras K, Thomson LE, Berman D, et al. Intracoronary cardiosphere-derived cells for heart regeneration after myocardial infarction (CADUCEUS): a prospective, randomised phase 1 trial. Lancet. 2012. DOI: 10.1016/S0140-6736(12)60195-0 35

[44] Siu CW, Tse HF. Cardiac regeneration: messages from CADUCEUS. Lancet. 2012. DOI: 10.1016/S0140-6736(12)60236-0 35

[45] Mirotsou M, Jayawardena TM, Schmeckpeper J, Gnecchi M, Dzau VJ. Paracrine mechanisms of stem cell reparative and regenerative actions in the heart. J Mol Cell Cardiol. 2011;50:280–9. DOI: 10.1016/j.yjmcc.2010.08.005 36

[46] Gnecchi M, Zhang Z, Ni A, Dzau VJ. Paracrine mechanisms in adult stem cell signaling and therapy. Circ Res. 2008;103:1204–19. DOI: 10.1161/CIRCRESAHA.108.176826 36

[47] Laflamme MA, Zbinden S, Epstein SE, Murry CE. Cell-based therapy for myocardial ischemia and infarction: pathophysiological mechanisms. Annu Rev Pathol. 2007;2:307–39. DOI: 10.1146/annurev.pathol.2.010506.092038 36

BIOMATERIALS – POLYMERS, SCAFFOLDS, AND BASIC DESIGN CRI...

CHAPTER 4

Biomaterials – Polymers, Scaffolds, and Basic Design Criteria

CHAPTER SUMMARY

Biomaterials constitute a major component in various strategies of tissue engineering and regeneration, as standalone treatments or in combination with cells and/or bioactive molecules. This chapter provides a brief summary of the "need to know" about biomaterials; such as the basic criteria for material selection and design, the type of natural and synthetic polymers in use, scaffold types, and their fabrication methodology. The goal here is to familiarize the readers with the basic terminology, concepts, and principles in the biomaterials research field as related to the tissue engineering strategy. The application of biomaterials in the different strategies of cardiac tissue engineering and regeneration will be described in the coming chapters.

4.1   INTRODUCTION

Tissue engineering aims at regenerating a living functional tissue to restore or establish the normal and original function of the damaged or compromised tissue [1]. As such, this multidisciplinary field implements knowledge and tools from diversified fields, such as material science, engineering, as well as cell and developmental biology. Biomaterials and scaffolds are critical components in this strategy, as standalones or in combination with cells and/or bioactive molecules (Fig. 4.1). They have been used as: 1) ECM replacement in a-cellular strategies; 2) controlled delivery systems for bioactive molecules; 3) vehicles for stem cell delivery and enhanced retention in the damaged tissue; and 4) supporting and guiding matrix for cell organization into a functional tissue, *in vitro* and *in vivo*.

This chapter provides an overview on biomaterials, their chemistry and sources, the basic criteria for their selection, and the methodologies for their fabrication as scaffolds.

4.2   BASIC BIOMATERIAL DESIGN CRITERIA

In general, the biomaterials used in each of the strategies for myocardial repair and regeneration (Fig. 4.1) should comply with the following basic criteria:

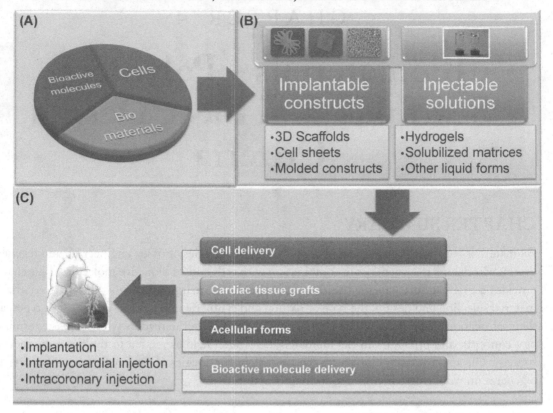

Figure 4.1: Paradigms for the use of biomaterials in cardiac tissue engineering and regeneration. Basic components (biomaterials, cells, and bioactive molecules) can be designed in various forms of implantable constructs or injectable solutions. These design strategies can be used for cell delivery, preparation of cardiac patches/grafts, *in situ* tissue support in acellular forms, or as a platform for the delivery of bioactive molecules. Ultimately, the biomaterial constructs or solutions are being delivered to the infarcted heart by implantation, or by using less invasive techniques, such as intramyocardial or intracoronary injections.

Biocompatibility – This term refers to the ability of a material to perform with an appropriate host response in a specific situation (Williams definition) [2]. In tissue engineering, biocompatibility refers to the ability of a scaffold to perform as a substrate that will support the appropriate cellular activity, including the facilitation of molecular and mechanical signaling, in order to optimize tissue regeneration, without eliciting any undesirable effects in those cells, or inducing any undesirable local or systemic responses in the host.

Mechanical strength –The strength of a material is its ability to withstand an applied stress without failure. Scaffolds in tissue engineering should have the mechanical properties to contain and protect the seeded or recruited cells and maintain their structure under mechanical perturbations existing during cultivation and at implant site. At the same time, the scaffold mechanical properties should be compatible with the host tissue to allow its integration without interfering with the normal function of the organ. This is especially critical when biomaterial is used as ECM replacement of damaged myocardium.

Biodegradation/bioresorption – Ideally, the scaffold should disappear from the host when tissue regeneration has been accomplished and normal function was restored. Biodegradable scaffolds can do so via polymer backbone degradation (e.g., hydrolysis, enzymatic cleavage) or by dissolution of the matrix. It is fundamental that the products of this process would be biocompatible and be resorbed by the body or removed from it via excretion from the urine.

Scaffold fabrication – Ideally, this process should be mild, using safe reagents and not affecting material properties, such as its cell recognition motifs. For example, cross-linking between polymer chains is often used in hydrogel fabrication from natural materials such as alginate, collagen, hyaluronan, and others. Cross-linking can be physical, where the polymer chains self-assemble due to electrostatic interactions, response to temperature and irradiation, or chemical, where covalent bonds are introduced between the polymer chains. Chemical cross-linking often changes the material properties (degradability, mechanical strength, and cell recognition) due to the lack of precise control over the position where the crosslink linkages are formed. In addition, chemical cross-linking often involves the use of harsh reagents, thus raising concerns about the material biocompatibility.

Scaffold internal morphology – When used as scaffolding for cells, the matrix should be porous with interconnecting pore structure and pore size larger than 50 μm, to enable cell-cell interactions and construct vascularization after implantation.

4.3 BIOMATERIAL CLASSIFICATION

Polymers commonly used for scaffold fabrication can be categorized by their source origin (natural or synthetic) and by their chemical structure (peptides/proteins, polysaccharides, polyesters, and others).

Natural polymers are usually biocompatible, easy to be chemically and physically modified and to be processed into various structures. Most of them, such as hyaluronan and collagen, possess cell recognition patterns enabling them to stimulate cell response. There are, however, concerns over their use in human therapy because of the risks of pathogen transmission and immune rejection associated with natural polymers that are produced from animal and cadaver sources.

Synthetic polymers can be designed with versatile properties, such as mechanical strength and biodegradation rate, and can be tailored with functional groups. A disadvantage that can be found in synthetic polymers, with respect to natural polymers, is the lack of the biological cues for promoting cell responses.

Table 4.1 lists the main materials used in cardiac tissue engineering strategies, as components in cellular constructs or standalone biomaterials treatments, and presents their main characteristics according to the design criteria listed above, namely, biocompatibility, mechanical strength, biodegradability/bioresorption, fabrication/cross-linking, and interactions with cells.

Table 4.1: Main biomaterials applied in cardiac tissue engineering [3, 4, 5, 6, 7, 8, 9]

Type	Material	Biodegradability/ control over process	Cross-linking	Mechanical strength	Cell adhesion	Biocompatibility/ Immunogenicity
Natural – proteins	Collagen	Biodegradable; poor	Chemical	Med-low	Good, cross-linking reduce	Med-immunogenicity after cross-linking
	Gelatin	Biodegradable; poor	Chemical	Low	Good	Low immunogenicity
	Fibrin	Biodegradable; good	Enzymatic	Med-low	Good	Non-immunogenic
Natural - polysaccharides	Alginate	Bioerodable matrix; good	Physical	Low	Poor	Non-immunogenic
	Chitosan	Biodegradable; good	Physical, chemical	Med-low	Poor	Low immunogenicity
	Hyaluronan	Biodegradable; poor	Chemical	Low	Good	Non-immunogenic
Synthetic	PIPAAm	Non-biodegradable; bioerodable	Chemical	NA	Good (temperature-dependent)	Biocompatible; NA
	Self-assembling peptides	Biodegradable; NA	Physical	NA	Good	Non-immunogenic
	Polyesters (PCL, PLA, PGA and copolymers)	Biodegradable; good	Chemical	High	Poor	Med-low-no immunogenicity

PIPAAm - poly (N-isopropylacrylamide); PCL- poly ε-caprolactone; PLA -polylactic acid; PGA-polyglycolic acid; NA – not available.

4.3.1   NATURAL PROTEINS

Collagen is a fibrous protein and the main component of ECM of mammalian tissues. About 25 types of collagen different in their chemical composition and molecular structure have been identified. Among the different collagen types, the fibrillar Type I collagen is the most abundant in nature and easy to produce. The biocompatibility, biodegradability, and cell-adhesive properties of the collagen type I matrix attributed to its selection by most researchers as the candidate scaffold for tissue growth and support. Collagen can be fabricated in many forms, such as hydrogel or macroporous scaffold; its fabrication frequently requires chemical cross-linking, which may affect its biological recognition by cells, biocompatibility, and degradability. Collagen is already commercialized as injectable product, thus it has been recognized as safe material by regulatory agencies. Two collagen-based products containing bone morphogenic protein (BMP)-2 or BMP-7 have been approved by the Food and

Drug Administration (FDA) in recent years for human clinical use: Infuse Bone Graft (Medtronik, US; Wyeth, UK), containing rhBMP-2, and Osigraft (Stryker Biotech) containing BMP-7.

In cardiac tissue engineering and myocardial repair after MI, collagen type I has been widely used as hydrogels or macroporous scaffolds for reconstructing the cardiac patch, either with or without seeded cells, as will be described in detail in upcoming chapters [8].

Gelatin (the irreversibly hydrolyzed form of collagen) has also been used in myocardial tissue engineering, especially in the format of a hydrogel prepared by chemical cross-linking. Gelatin is a biodegradable material, but under various conditions it can provoke an unspecific inflammatory response [10].

Fibrin(ogen) is a natural scaffold protein that is used extensively for medical applications (e.g., surgical adhesive sealant) and for myocardial tissue engineering. It is FDA-approved due to its favorable wound healing properties. Fibrinogen circulates as an inactive precursor in circulation and is recruited to injury sites where it becomes activated by proteolytic cleavage; the covalent cross-linking of fibrin by thrombin and factor XIIIa form the fibrin clot, a complex network, composed of fibrils. Fibrin clots provide a natural wound healing matrix that can be remodeled via cellular activities to form the tissue-specific mature ECM. Fibrin scaffolds were prepared as hydrogels and in injectable form, wherein its components (fibrinogen and thrombin) are mixed during injection into the tissue. Their application as 3D scaffolds for *in-vitro* cell culture is limited in time (4–7 days) due to their rapid degradation by enzymes secreted by the seeded cells. *In vivo*, they have been shown to increase cell retention after transplantation into infarcted hearts and in some studies, fibrin alone caused improvement in cardiac function, as will be described in Chapter 9 [8].

4.3.2 NATURAL POLYSACCHARIDES

Alginate is an anionic polysaccharide extracted from brown algae, composed of $1{\rightarrow}4$ linked β-D-mannuronic acid (M) and α-L-guluronic acid (G) (Fig. 4.2). Divalent cations, such as calcium ions, interact with high affinity with the G monomer blocks to form ionic bridges between different alginate chains ("egg box" model) eventually leading to hydrogel formation (Fig. 4.2).

The physical cross-linking of alginate represents a significant advantage, as the use of various chemical agents for gelation is eliminated. Since there is no known mammalian enzyme which degrades the alginate backbone, it is assumed that alginate is not degradable in mammals. Yet, the calcium-cross-linked hydrogel is readily erodible with time due to exchange of calcium ions by sodium ions in physiological milieu, leading to hydrogel dissolution. The water-soluble alginate chains are excreted through the kidney if the molecular weight is below 50 kDa [11].

The alginate matrix is inert and resistant to cell adhesion, thus it can be used as a "blank canvas" to investigate specific cell-matrix interactions by attaching biological cues, as will be described in detail in Chapter 6. The long experience with alginate as cell matrix and implant indicates its biocompatibility.

Chitosan is a biodegradable cationic polysaccharide made of D-glucosamine and N-acetyl-D-glucosamine linked by $\beta(1,4)$ glycoside bonds. Due to its positive charge, it can ionically interact with

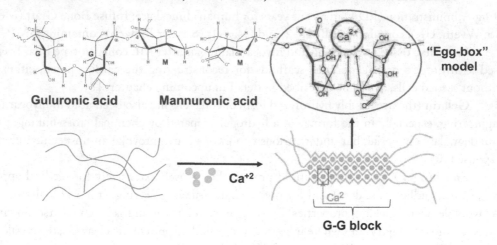

Figure 4.2: Alginate structure and the egg-box model of hydrogel formation.

negatively charged polymers and/or ions and molecules. Chitosan is abundant in nature; it derives from the deacetylation of chitin that forms the crustaceous skeleton. The polymer is soluble in water at acidic pH, but it can be chemically modified and/or salified to make it soluble in physiological pH. It is degraded by enzymes, such as lysozymes. Due to its origin, a matrix made of chitosan does not support specific and high affinity interactions with mammalian cells, and attachments of recognition peptide to the polymer are required to enable such interactions. Scaffolds made of chitosan and/or its modified forms have been applied for cardiac tissue engineering, as further described in Chapter 9.

Hyaluronan (HA) is a linear un-sulfated polysaccharide, composed of repeating disaccharides $[(1\rightarrow3)\text{-}\beta\text{-N-acetyl-D-glucosamine-}(1\rightarrow4)\text{-}\beta\text{-D-glucoronic acid}]$. In the human body, HA is found primarily in the extracellular and pericellular matrix; its degradation occurs by hyaluronidases. The HA has versatile biological functions, such as a lubricant material and numerous receptor-mediated roles in different cell processes. The processing of HA as scaffolds for tissue engineering requires chemical modification of the material to achieve cross-linking, such as by photopolymerization. The un-cross-linked HA is not effective as an injectable material due to its poor mechanical properties, rapid degradation, and clearance *in vivo* [12, 13].

4.3.3   SYNTHETIC PEPTIDES AND POLYMERS

Among the recently developed synthetic biomaterials, self-assembling peptides have become a favorable option as scaffold materials since they are composed of natural building blocks, they can be synthesized with the appropriate features, and the manufacturing of scaffolds from them is simple and mild. These peptides are usually 8-16 amino acids long and composed of alternating hydrophobic and hydrophilic residues. They form stable β-sheets in water, and upon exposure to physiological salt concentrations or pH they form a stable hydrogel made of flexible nanofibers. Such peptide nanofibers have been shown to create a favorable cell microenvironment upon injection into the infarct, and as systems for growth factor delivery [3] (details in Chapter 10).

Poly (N-isopropylacrylamide) (PIPAAm) has been mainly applied for the creation of intact cell sheets. When PIPAAm is grafted onto tissue culture plates by electron beam irradiation, the grafted surfaces become slightly hydrophobic under cell culture condition, at 37°C, but readily become hydrated and hydrophilic below the polymer LCST (low critical solution temperature of PIPAAm), 32°C. Thus, the attachment and detachment of cells on the culture surface can be controlled by simple temperature change. With the temperature-responsive culture surfaces, cells can be non-invasively harvested as intact cell sheets along with their deposited ECM, and subsequently be attached to host tissue. Using this technique, several cell sheets could be layered on top of each other to create multilayered three-dimensional cell constructs [14, 15].

Other types of more traditional synthetic polymers that have been used in tissue engineering, but not as much in cardiac tissue engineering, especially in their injectable forms include: the biodegradable poly(α-hydroxyacid) [poly ε-caprolactone (PCL), polylactic acid (PLA), polyglycolic acid] and their copolymers [10, 16, 17, 18]. These polymers were adapted from the field of drug delivery systems and the main initial motivation for their use in tissue engineering stems from their biodegradability, biocompatibility, and FDA approval. These polymers have been mainly used as solid macroporous scaffolding for cardiac patch reconstruction.

4.4   BASIC SCAFFOLD FABRICATION FORMS

Scaffolds can be fabricated in different shapes, sizes, and internal structure (porosity, pore size, and architecture of the pore structure (isotropic or anisotropic). In tissue engineering of patches, either with seeded cells or without cells, the two main scaffolds in use are hydrogels and macro-porous solid scaffolds.

4.4.1   HYDROGELS

Hydrogel is a network of polymer chains that are water-insoluble, sometimes found as a colloidal gel in which water is the dispersion medium. Hydrogels are superabsorbent (they can contain over 99% water) natural or synthetic polymers. Hydrogels also possess a degree of flexibility very similar to natural ECM, due to their significant water content. The hydrogels can be prepared from natural and synthetic polymers by physical/ionic interactions (alginate) or via chemical cross-linking (collagen, HA, and others). Cells are incorporated/encapsulated in the hydrogel during fabrication. Due to their resemblance to ECM texture, hydrogels are extensively being investigated as ECM replacements for damaged ECM after MI. They are delivered either by intramyocardial injection or by catheter-based techniques via the intracoronary route [19, 20, 21, 22, 23, 24].

4.4.2   MACROPOROUS SCAFFOLDS

Macroporous scaffolds are characterized by large pore size (50-200 μm in diameter) and matrix porosity (70-90%). The pore size in scaffolds should be at least 50 μm in diameter to enable vascularization (blood vessel penetration) after their implantation. The pore size and architecture as well as the extent of pore interconnectivity are major effectors on cell seeding, cell penetration from the host, and cell organization into a tissue. The most common techniques for preparing macroporous scaffolds are: solvent casting particulate (porogen) leaching, non-solvent induced phase separation, thermally induced phase separation, foaming process, microsphere sintering, and electrospinning.

Recently proposed fabrication techniques are: rapid prototyping, solid free-form, shape deposition manufacturing, fused deposition modeling, 3D printing, selective laser sintering, stereolitographic technique, and molecular self-assembly [25].

Macroporous alginate scaffold, commercially available from Life Technologies Incorporation as AlgiMatrix^{TM}, has been developed by our group, by a controlled freeze-dry technique of calcium cross-linked alginate solution [26, 27]. The scaffold porous structure was dependent on the freezing regime (rate and direction) (Fig. 4.3A, B). When the calcium crosslinked alginate solutions were slowly frozen at $-20°$C, in a nearly homogenous cold atmosphere, the resultant scaffold had an isotropic pore structure; the pores were spherical and interconnected. In contrast, when the cooling process was performed under the unidirectional temperature gradient along the freezing solution, an anisotropic pore structure was attained. This pore architecture influenced the shape of the tissue organized in the different scaffolds (Fig. 4.3C-E).

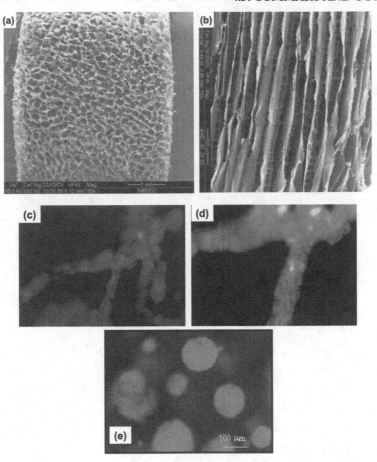

Figure 4.3: Alginate scaffold porosity affects cell behavior and tissue morphology. Depending on the freezing regime, the scaffolds can be prepared with isotropic (**A**) or anisotropic (**B**) pore structure. The pore architecture influences cell organization in the scaffold: endothelial cells cultivated in anisotropic alginate scaffolds (**C**, magnified in **D**), and C3A (human hepatocyte cell line) spheroids grown in isotropic alginate scaffolds (**E**).

4.5 SUMMARY AND CONCLUSIONS

This chapter provided an overview of the biomaterials used in tissue engineering. It presented the basic criteria for material selection and design, the type of natural and synthetic polymers in use and their advantages/drawbacks, as well as scaffold types and their fabrication methodology. The summary is not exhaustive but focuses on those concepts that have given or are expected to give significant input to a better understanding of the biomaterials and their application in various strategies of cardiac tissue engineering and regeneration.

BIBLIOGRAPHY

[1] Langer R, Vacanti JP. Tissue engineering. Science. 1993;260:920–6.
DOI: 10.1126/science.8493529 41

[2] Williams DF. The Williams Dictionary of Biomaterials: Liverpool University Press; 1999. 42

[3] Segers VF, Lee RT. Local delivery of proteins and the use of self-assembling peptides. Drug
discovery today. 2007;12:561–8. DOI: 10.1016/j.drudis.2007.05.003 44, 47

[4] Pok S, Jacot JG. Biomaterials advances in patches for congenital heart defect repair. J Cardiovasc
Transl Res. 2011;4:646–54. DOI: 10.1007/s12265-011-9289-8 44

[5] Haraguchi Y, Shimizu T, Yamato M, Okano T. Regenerative therapies using cell sheet-based
tissue engineering for cardiac disease. Cardiol Res Pract. 2011;2011:845170.
DOI: 10.4061/2011/845170 44

[6] Drury JL, Dennis RG, Mooney DJ. The tensile properties of alginate hydrogels. Biomaterials.
2004;25:3187–99. DOI: 10.1016/j.biomaterials.2003.10.002 44

[7] Tous E, Purcell B, Ifkovits JL, Burdick JA. Injectable acellular hydrogels for cardiac repair. J
Cardiovasc Transl Res. 2011;4:528–42. DOI: 10.1007/s12265-011-9291-1 44

[8] Venugopal JR, Prabhakaran MP, Mukherjee S, Ravichandran R, Dan K, Ramakrishna S. Biomaterial strategies for alleviation of myocardial infarction. J R Soc Interface. 2012;9:1–19.
DOI: 10.1098/rsif.2011.0301 44, 45

[9] Allison DD, Grande-Allen KJ. Review. Hyaluronan: a powerful tissue engineering tool. Tissue
Eng. 2006;12:2131–40. DOI: 10.1089/ten.2006.12.2131 44

[10] Akhyari P, Kamiya H, Haverich A, Karck M, Lichtenberg A. Myocardial tissue engineering:
the extracellular matrix. Eur J Cardiothorac Surg. 2008;34:229–41.
DOI: 10.1016/j.ejcts.2008.03.062 45, 47

[11] Al-Shamkhani A, Duncan, R. Radioiodination of alginate via covalently-bound tyrosinamide
allows monitoring of its fate in vivo. Journal of Bioactive and Compatible Polymers. 1995;10:4–
13. DOI: 10.1177/088391159501000102 45

[12] Prestwich GD, Kuo JW. Chemically-modified HA for therapy and regenerative medicine.
Current pharmaceutical biotechnology. 2008;9:242–5. DOI: 10.2174/138920108785161523
46

[13] Prestwich GD. Hyaluronic acid-based clinical biomaterials derived for cell and molecule delivery in regenerative medicine. J Control Release. 2011;155:193–9.
DOI: 10.1016/j.jconrel.2011.04.007 46

[14] Masuda S, Shimizu T, Yamato M, Okano T. Cell sheet engineering for heart tissue repair. Adv Drug Deliv Rev. 2008;60:277–85. DOI: 10.1016/j.addr.2007.08.031 47

[15] Shimizu T, Yamato M, Kikuchi A, Okano T. Cell sheet engineering for myocardial tissue reconstruction. Biomaterials. 2003;24:2309–16. DOI: 10.1016/S0142-9612(03)00110-8 47

[16] Giraud MN, Armbruster C, Carrel T, Tevaearai HT. Current state of the art in myocardial tissue engineering. Tissue Eng. 2007;13:1825–36. DOI: 10.1089/ten.2006.0110 47

[17] Wang F, Guan J. Cellular cardiomyoplasty and cardiac tissue engineering for myocardial therapy. Adv Drug Deliv Rev. 2010;62:784–97. DOI: 10.1016/j.addr.2010.03.001 47

[18] Jawad H, Ali NN, Lyon AR, Chen QZ, Harding SE, Boccaccini AR. Myocardial tissue engineering: a review. J Tissue Eng Regen Med. 2007;1:327–42. DOI: 10.1161/01.RES.0000196562.73231.7d 47

[19] Christman KL, Fok HH, Sievers RE, Fang Q, Lee RJ. Fibrin glue alone and skeletal myoblasts in a fibrin scaffold preserve cardiac function after myocardial infarction. Tissue Eng. 2004;10:403–9. DOI: 10.1089/107632704323061762 47

[20] Christman KL, Vardanian AJ, Fang Q, Sievers RE, Fok HH, Lee RJ. Injectable fibrin scaffold improves cell transplant survival, reduces infarct expansion, and induces neovasculature formation in ischemic myocardium. J Am Coll Cardiol. 2004;44:654–60. DOI: 10.1016/j.jacc.2004.04.040 47

[21] Dai W, Wold LE, Dow JS, Kloner RA. Thickening of the infarcted wall by collagen injection improves left ventricular function in rats: a novel approach to preserve cardiac function after myocardial infarction. J Am Coll Cardiol. 2005;46:714–9. DOI: 10.1016/j.jacc.2005.04.056 47

[22] Landa N, Miller L, Feinberg MS, Holbova R, Shachar M, Freeman I, et al. Effect of injectable alginate implant on cardiac remodeling and function after recent and old infarcts in rat. Circulation. 2008;117:1388–96. DOI: 10.1161/CIRCULATIONAHA.107.727420 47

[23] Leor J, Tuvia S, Guetta V, Manczur F, Castel D, Willenz U, et al. Intracoronary injection of in situ forming alginate hydrogel reverses left ventricular remodeling after myocardial infarction in Swine. J Am Coll Cardiol. 2009;54:1014–23. DOI: 10.1016/j.jacc.2009.06.010 47

[24] Wang T, Wu DQ, Jiang XJ, Zhang XZ, Li XY, Zhang JF, et al. Novel thermosensitive hydrogel injection inhibits post-infarct ventricle remodelling. Eur J Heart Fail. 2009;11:14–9. DOI: 10.1093/eurjhf/hfn009 47

[25] Weigel T, Schinkel G, Lendlein A. Design and preparation of polymeric scaffolds for tissue engineering. Expert review of medical devices. 2006;3:835–51. DOI: 10.1586/17434440.3.6.835 48

[26] Shapiro L, Cohen S. Novel alginate sponges for cell culture and transplantation. Biomaterials. 1997;18:583–90. DOI: 10.1016/S0142-9612(96)00181-0 48

[27] Zmora S, Glicklis R, Cohen S. Tailoring the pore architecture in 3-D alginate scaffolds by controlling the freezing regime during fabrication. Biomaterials. 2002;23:4087–94. DOI: 10.1016/S0142-9612(02)00146-1 48

CHAPTER 5

Biomaterials as Vehicles for Stem Cell Delivery and Retention in the Infarct

CHAPTER SUMMARY

Poor cell engraftment and survival in the infarct are the major limitations of the strategy is of cell suspension transplantation to treat and regenerate the infarcted myocardium after MI. This chapter describes the application of biomaterials as delivery vehicles to improve cell survival and function after transplantation. The results of this strategy, in terms of cell retention, integration, and beneficial effect on cardiac repair are presented for the different stem and progenitor cells used in cardiac repair. At the end, we present the "MAGNUM" phase 1 clinical trial with implantable cardiac patches based on bone marrow cells seeded in collagen type I that provided an initial proof-of-concept for the potential use of biomaterials to enhance cell integration and cardiac repair.

5.1   INTRODUCTION

As already mentioned, a large body of evidence from preclinical and clinical trials indicates some functional improvements in heart function after MI, even with the injections of suspensions of non-contractile cells. These include various populations of stem/progenitor cells (bone marrow, adipose, or cardiac tissue-derived). Aside from the apparent efficacy shown by stem cell transplantation in various animal models, several critical hurdles associated with this strategy arose. First, the retention of cells immediately after delivery is highly dependent on the administration strategy: if cells are injected intramyocardially during open-chest surgery, many cells are lost through the vasculature, and only a few cells infused into the coronary arteries do ultimately engraft. Second, survival in the inflammatory environment of the acute infarcted myocardium is a challenge common to all types of transplanted cells, as typically 90% of the cells die within a week after transplantation. In addition, cell retention is extremely variable from one study to another, making final graft size unpredictable. Finally, virtually all studies involving cell suspension transplantation into the heart have found that the scar tissue forms a major barrier to proper integration of the implanted cells [1, 2].

The use of biomaterials as a platform or vehicle for cell delivery into the infarcted myocardium seems like a very logical strategy, which can potentially reduce or eliminate the obstacles seen in cell suspension transplantation [3]. The biomaterials can protect the implanted cells from the aggressive environment after acute MI. With proper selection or design, the biomaterials can serve as an artificial extracellular matrix, to replace the degraded ECM after MI and provide the required temporal support to allow cell engraftment and retention. Using biomaterial scaffolds, thicker cell constructs could be created *in vitro*, to achieve high density of cells in future grafts. In addition, biomaterials could provide various biochemical and biophysical cues to enhance cell survival, induce angiogenesis at infarct, and direct and control stem/progenitor cell differentiation, by implementation of controlled delivery mechanisms for growth factors, integration of specific cell-matrix interactions, and more. The successful implementation of biomaterials as cell delivery vehicle promises to yield an improved cell engraftment and long-term functionality.

5.2   STEM CELL DELIVERY BY BIOMATERIALS

Table 5.1 summarizes the results of several studies using biomaterials as vehicles for stem cell delivery in the infarct.

5.2.1   HUMAN EMBRYONIC STEM CELL-DERIVED CELLS

A complimentary approach to the use of human ESC-derived cardiomyocytes for preservation of myocardial tissue after MI is the rescue of the vascular network compromised during ischemia insult. Myocardial injection of hESC-derived vascular cells in an *in situ* formed bioactive (thymosin β4-containing) PEG hydrogel into infarcted hearts in rats has shown that the delivered cells formed capillaries in the infarct zone. In addition, magnetic resonance imaging (MRI) revealed that the microvascular grafts effectively preserved contractile performance, attenuated left ventricular dilation, and decreased infarct size [4]. In another study, implantation of porous fibrin scaffold co-seeded with hESC-derived endothelial and smooth muscle cells, in a porcine model of ischemia/reperfusion resulted in significant LV functional improvement as judged by cardiac MRI. The authors point to neovascularization as an underlying mechanism behind function restoration [5].

Although the latter study confirmed improvements in cell engraftment due the delivery in biomaterial scaffolds (shown by bioluminescent imaging of luciferase-labeled cells after four weeks), both studies did not provide a direct measure of the effect of scaffolds on the extent of cell retention, as the control group of direct cell injection was missing. Nevertheless, in addition to some beneficial effect of sole material, the authors point to neovascularization as a major paracrine mechanism responsible for functional repair by the delivered cells.

5.2.2   ADULT BONE MARROW-DERIVED STEM CELLS

To date, a handful of studies have attempted at regenerating the ischemic myocardium via implantation of a tissue patch pre-seeded either with bone marrow cells or BM-derived mesenchymal stem cells.

Piao *et al* used bone marrow-derived mononuclear cell (BMMNC)-seeded biodegradable poly-glycolide-co-caprolactone (PGCL) scaffolds in a rat MI model. Patch implantation resulted

Table 5.1: Selected biomaterial-based strategies for stem cell delivery

Cell type	Biomaterial and form	Major mechanism of improvement	Reference
Non-autologous			
ESC-derived cells			
Vascular cells	PEG hydrogel +Tβ4	Retention Angiogenesis	[4]
Endothelial/SMC	Fibrin scaffold	Angiogenesis	[5]
Autologous			
EPC	Vitronectin/collagen scaffold+SDF-1	Angiogenesis	[6]
Cardiac progenitor cells	Self-assembled peptides+IGF-1	Cardiomyogenesis Angiogenesis	[7]
Adipose tissue-derived stem cells			
MSC	Cell sheets (on PIPAAm)	Retention Angiogenesis	[8]
MSC	Fibrin or collagen hydrogels	Retention	[9]
Bone marrow-derived stem cells			
MNC	Self-assembled peptides	Retention Angiogenesis	[10]
MNC	PGCL scaffold	Retention Angiogenesis	[11]
MSC	Collagen hydrogel	NA	[12]
MSC	PLCL scaffold	Retention Infarct size	[13]
MSC	RGD-alginate microspheres	Retention Angiogenesis	[14]
MSC	Pullulan/dextran scaffold	Retention Angiogenesis	[15]

PEG- polyethyleneglycol; Tβ4 – thymosin β4; SMC- smooth muscle cells; EPC – endothelial progenitor cells; SDF-1 –stromal cell-derived factor-1; IGF-1 – insulin-like growth factor-1; PIPAAm – poly-N-isopropylacrylamide; MNC – mononuclear cells; PGCL - poly-glycolide-co-caprolactone; MSC – mesenchymal stem cells; PLCL- poly(lactide-co-ε-caprolactone).

in improved neovascularization and the presence of α-myosin heavy chain (MHC) and troponin I markers in some BMMNCs. Interestingly, when either BMMNC- and acellular-PGCL patches were used, attenuation of LV remodeling and LV dysfunction was observed, suggesting the potential of empty scaffolding as an effective treatment alternative for MI [11]. Lin *et al* used self-assembling peptide nanofibers (NFs) for BMMNC delivery in a pig MI model. NF injection significantly improved diastolic function and reduced ventricular remodeling 28 days after treatment. Injection of BMMNCs alone ameliorated systolic function only, whereas the injection of BMMNCs with NFs significantly improved both systolic and diastolic functions, increased transplanted cell retention ($\sim$30 cells/mm^2 in BMMNC-only group and $\sim$230 cells/mm^2 in BMMNCs/NFs-treated animals) and increased capillary density in the peri-infarct area. The authors indicated an improved cell retention, survival and function, together with a synergistic effect of using biomaterials, as the major mechanisms behind the observed beneficial results [10].

Kim and co-workers showed that epicardial implantation of a poly(lactide-co-ε-caprolactone) (PLCL) patch seeded with MSC induced an enhanced expression of cardiac markers such as MHC, α-actin and troponin I as well as the cardiac transcription factor GATA-4 when compared to MSC suspension injections. Though MSC injections significantly improved heart function, the MSC-PLC patch had a greater positive effect on LV EF and infarct size. However, the likely possibility that these beneficial effects, at least partially, stem from greater cell retention in PLC scaffolds, has not been elaborated, as the retention parameter was not compared between the cell-treated groups [13]. Chen *et al* developed bioengineered tissue graft, by using a porous acellular bovine pericardium sandwiched with multilayered sheets of mesenchymal stromal cells. This tissue graft (sandwiched patch) was used to replace the resected infarct scar area in a syngeneic Lewis rat model with an experimentally chronic MI. Patch application one month after MI improved cardiac function, compared to empty patch-treated animals. The effect of the biomaterial on cell retention was not evaluated [16].

Yu *et al* examined the effect of encapsulation of human MSCs (hMSCs) in RGD-modified alginate hydrogel microspheres on cell retention, differentiation, and myocardial repair in rats. The encapsulation significantly improved cell survival, compared to simple cell injection. Specifically, hMSC presence (presented as percentage of human cells to host (rat) cells, quantified by real-time PCR), at seven days post-injection in immunodeficient rats, was evident only in the hMSCs-encapsulated group (0.58%). After an additional week, the cells in the microbead group still indicated good retention (0.53%). There were no significant differences in angiogenesis degree between the treatment groups; all showed higher vessel density compared to untreated controls. Alginate microspheres, with or without encapsulated MSCs, successfully maintained LV shape and prevented negative LV remodeling after MI, emphasizing again the importance of the biomaterial in preserving mechanical and passive properties of the myocardium [14].

In a recent study, Norol and co-workers directly compared the engraftment rates of rat MSC when delivered within porous pullulan/dextran scaffold or endocardial injection of their cell suspension in a rat model of MI. Cellular engraftment was measured by quantitative RT-PCR using

MSCs transduced with GFP. The use of a scaffold promoted local cellular engraftment and survival. The number of residual GFP^+ cells was greater with the scaffold than after cell suspension injection (9.7% vs. 5.1% at one month and 16.3% vs. 6.1% at two months, respectively). This concurred with a significant increase in mRNA VEGF level in the scaffold group. Clusters of GFP^+ cells were detected in the peri-infarct area, mainly phenotypically consistent with immature MSCs. Functional assessment by echocardiography at two months post-infarct also showed a trend toward a lower left ventricular dilatation and a reduced fibrosis in the scaffold group in comparison to direct cell injection group [15].

5.2.3   ADULT ADIPOSE TISSUE-DERIVED STEM CELLS

Adipose tissue is an appealing source of MSCs for clinical autologous cell therapies and tissue engineering [17]. Miyahara *et al* used the cell sheet technology (see Section 6.3 for more details) to regenerate the infarcted myocardium with autologous adipose tissue-derived MSCs in rats with chronic heart failure secondary to MI. *In situ* growth of the adipose tissue-derived monolayered MSCs developed into neovessel-rich stratum that also contained some undifferentiated cells and cardiomyocytes. The engrafted MSC tissue improved heart function and reversed wall thinning of the infarcted area when compared to control animals treated with a monolayer of dermal fibroblasts. The restorative effect of the stem cell monolayer was attributed to the MSC capacity to differentiate into vascular cells and to secrete angiogenic cytokines [8]. Hamdi *et al* used autologous adipose stem cell sheets for MI repair in rats. Epicardial deposition of cell sheets resulted in increased cell survival and engraftment (shown by larger number of eGFP-positive cells in the epicardial and intramyocardial layers), associated with better preservation of LV geometry, when compared to dissociated cell injection [18].

5.2.4   CARDIAC STEM/PROGENITOR CELLS

Although several studies are already performed using direct injection of various CSC/CPC types in MI models, tissue engineering approaches utilizing these cells are still in their infancy. Zakharova *et al* used a cardiosphere-derived cell sheet approach to restore infarcted myocardium in a rat model. Delivery of rat or human cardiac progenitor cells as cell sheets had a very positive effect on cell survival and facilitated migration from the sheet to the area of ischemic injury. Moreover, cardiosphere-derived cell sheet transplantation promoted cardiomyogenic and vascular differentiation, and significantly reduced adverse LV remodeling [19].

5.3   CLINICAL TRIALS

In the MAGNUM (Myocardial Assistance by Grafting a New Bioartificial Upgraded Myocardium) phase 1 clinical trial, bone marrow cells seeded in collagen type 1 patch were implanted onto 10 patients who had coronary artery bypass grafts and were intramyocardially injected with suspensions of autologous bone marrow cells. In 10 other patients, the treatment was only cell suspension injection into the infarcted myocardium. The results of this investigation revealed that the combined strategy of cell suspension injection and implantation of the cardiac patch on the infarct resulted in greater LV ED volume and scar thickness compared to cell suspension injection alone, at a 10±3.5 months follow-up. Both treatments improved ejection fraction in those patients compared with the baseline measurements performed one week before coronary artery bypass [20, 21]. Clearly, the implantation of biomaterial scaffolds onto the infarct zone significantly contributed to the maintenance of LV wall thickness, probably by increasing the retention of transplanted cells as well as by providing space and support for penetrating host cells. Together, this resulted in thicker scars and reduced stress on LV wall, eventually attenuating remodeling of the heart.

5.4 SUMMARY AND CONCLUSIONS

This chapter presented a handful of studies aimed to investigate the applications of biomaterials as stem cell delivery vehicles. Only a few of these studies directly measured the extent of cell retention and engraftment. These studies show increased percentage of cells surviving in the infarct, when delivered as a biomaterial patch. However, the long-term engraftment (several months) is still low (~10-20%), emphasizing the need for continuous refinement and engineering of the biomaterial schemes, and the possible introduction of pro-survival factors. In parallel, the feasibility and efficacy of biomaterial-based cell delivery awaits confirmation in large and randomized clinical trials.

BIBLIOGRAPHY

[1] Laflamme MA, Murry CE. Regenerating the heart. Nat Biotechnol. 2005;23:845–56. DOI: 10.1038/nbt1117 55

[2] Segers VF, Lee RT. Stem-cell therapy for cardiac disease. Nature. 2008;451:937–42. DOI: 10.1038/nature06800 55

[3] Segers VF, Lee RT. Biomaterials to enhance stem cell function in the heart. Circ Res. 2011;109:910–22. DOI: 10.1161/CIRCRESAHA.111.249052 55

[4] Kraehenbuehl TP, Ferreira LS, Hayward AM, Nahrendorf M, van der Vlies AJ, Vasile E, et al. Human embryonic stem cell-derived microvascular grafts for cardiac tissue preservation after myocardial infarction. Biomaterials. 2011;32:1102–9. DOI: 10.1016/j.biomaterials.2010.10.005 56

[5] Xiong Q, Hill KL, Li Q, Suntharalingam P, Mansoor A, Wang X, et al. A fibrin patch-based enhanced delivery of human embryonic stem cell-derived vascular cell transplantation in a porcine model of postinfarction left ventricular remodeling. Stem Cells. 2011;29:367–75. DOI: 10.1002/stem.580 56

[6] Frederick JR, Fitzpatrick JR, 3rd, McCormick RC, Harris DA, Kim AY, Muenzer JR, et al. Stromal cell-derived factor-1alpha activation of tissue-engineered endothelial progenitor cell

matrix enhances ventricular function after myocardial infarction by inducing neovasculogenesis. Circulation. 2010;122:S107–17. DOI: 10.1161/CIRCULATIONAHA.109.930404

[7] Padin-Iruegas ME, Misao Y, Davis ME, Segers VF, Esposito G, Tokunou T, et al. Cardiac progenitor cells and biotinylated insulin-like growth factor-1 nanofibers improve endogenous and exogenous myocardial regeneration after infarction. Circulation. 2009;120:876–87. DOI: 10.1161/CIRCULATIONAHA.109.852285

[8] Miyahara Y, Nagaya N, Kataoka M, Yanagawa B, Tanaka K, Hao H, et al. Monolayered mesenchymal stem cells repair scarred myocardium after myocardial infarction. Nature medicine. 2006;12:459–65. DOI: 10.1038/nm1391 59

[9] Danoviz ME, Nakamuta JS, Marques FL, dos Santos L, Alvarenga EC, dos Santos AA, et al. Rat adipose tissue-derived stem cells transplantation attenuates cardiac dysfunction post infarction and biopolymers enhance cell retention. PLoS One. 2010;5:e12077. DOI: 10.1371/journal.pone.0012077

[10] Lin YD, Yeh ML, Yang YJ, Tsai DC, Chu TY, Shih YY, et al. Intramyocardial peptide nanofiber injection improves postinfarction ventricular remodeling and efficacy of bone marrow cell therapy in pigs. Circulation. 2010;122:S132–41. DOI: 10.1161/CIRCULATIONAHA.110.939512 58

[11] Piao H, Kwon JS, Piao S, Sohn JH, Lee YS, Bae JW, et al. Effects of cardiac patches engineered with bone marrow-derived mononuclear cells and PGCL scaffolds in a rat myocardial infarction model. Biomaterials. 2007;28:641–9. DOI: 10.1016/j.biomaterials.2006.09.009 58

[12] Simpson D, Liu H, Fan TH, Nerem R, Dudley SC, Jr. A tissue engineering approach to progenitor cell delivery results in significant cell engraftment and improved myocardial remodeling. Stem Cells. 2007;25:2350–7. DOI: 10.1634/stemcells.2007-0132

[13] Jin J, Jeong SI, Shin YM, Lim KS, Shin H, Lee YM, et al. Transplantation of mesenchymal stem cells within a poly(lactide-co-epsilon-caprolactone) scaffold improves cardiac function in a rat myocardial infarction model. Eur J Heart Fail. 2009;11:147–53. DOI: 10.1093/eurjhf/hfn017 58

[14] Yu J, Du KT, Fang Q, Gu Y, Mihardja SS, Sievers RE, et al. The use of human mesenchymal stem cells encapsulated in RGD modified alginate microspheres in the repair of myocardial infarction in the rat. Biomaterials. 2010;31:7012–20. DOI: 10.1016/j.biomaterials.2010.05.078 58

[15] Visage CL, Gournay O, Benguirat N, Hamidi S, Chaussumier L, Mougenot N, et al. Mesenchymal stem cell delivery into rat infarcted myocardium using a porous polysaccharide-based scaffold: a quantitative comparison with endocardial injection. Tissue Eng Part A. 2012;18:35–44. DOI: 10.1089/ten.tea.2011.0053 59

[16] Chen CH, Wei HJ, Lin WW, Chiu I, Hwang SM, Wang CC, et al. Porous tissue grafts sandwiched with multilayered mesenchymal stromal cell sheets induce tissue regeneration for cardiac repair. Cardiovasc Res. 2008;80:88–95. DOI: 10.1093/cvr/cvn149 58

[17] Mazo M, Gavira JJ, Pelacho B, Prosper F. Adipose-derived stem cells for myocardial infarction. J Cardiovasc Transl Res. 2011;4:145–53. DOI: 10.1007/s12265-010-9246-y 59

[18] Hamdi H, Planat-Benard V, Bel A, Puymirat E, Geha R, Pidial L, et al. Epicardial adipose stem cell sheets results in greater post-infarction survival than intramyocardial injections. Cardiovasc Res. 2011;91:483–91. DOI: 10.1093/cvr/cvr099 59

[19] Zakharova L, Mastroeni D, Mutlu N, Molina M, Goldman S, Diethrich E, et al. Transplantation of cardiac progenitor cell sheet onto infarcted heart promotes cardiogenesis and improves function. Cardiovasc Res. 2010;87:40–9. DOI: 10.1093/cvr/cvq027 59

[20] Chachques JC, Trainini JC, Lago N, Masoli OH, Barisani JL, Cortes-Morichetti M, et al. Myocardial assistance by grafting a new bioartificial upgraded myocardium (MAGNUM clinical trial): one year follow-up. Cell Transplant. 2007;16:927–34. DOI: 10.3727/096368907783338217 60

[21] Chachques JC, Trainini JC, Lago N, Cortes-Morichetti M, Schussler O, Carpentier A. Myocardial Assistance by Grafting a New Bioartificial Upgraded Myocardium (MAGNUM trial): clinical feasibility study. Ann Thorac Surg. 2008;85:901–8. DOI: 10.1016/j.athoracsur.2007.10.052 60

CHAPTER 6

Bioengineering of Cardiac Patches, *In Vitro*

CHAPTER SUMMARY

In vitro-generated cardiac tissue grafts represent an ideal solution when the replacement of significant portions of the myocardium (damaged by myocardial infarction or bearing a structural defect) is required. This chapter introduces the principles of *in vitro* cardiac tissue engineering and presents three main strategies developed to attain cardiac patches, *in vitro*: cell entrapment in hydrogels, cell sheet technology, and cell seeding in preformed macroporous scaffolds. It describes the various materials and scaffolds in use, the evolution of new synthetic scaffolds by integration of cell-matrix interactions bio-inspired by ECM to improve tissue organization and introduces recent advancements in micro- and nano-fabrication techniques applied for the fabrication of scaffolds bio-mimicking the physical structure and architecture of cardiac ECM. In Chapter 7, we will present the dynamic environment created by perfusion bioreactors and stimulation patterns, which is necessary for attaining a functional cardiac patch.

6.1 INTRODUCTION

The replacement of a large scar tissue after MI and the corrections of congenital heart malformations, such as septal defects, require the transplantation of a fully developed cardiac tissue graft prepared *in vitro*. The engineered cardiac patch should be thick and has to display the functional and morphological properties of the native cardiac muscle. Once integrated into the heart, the cardiac patch must develop systolic force, withstand diastolic load with appropriate compliance, and form an electrical and functional syncytium with the host myocardium [1].

Three main strategies for *in vitro* cardiac patch reconstruction have been developed over the last decade or so and they are based on seeding the cardiac cells in scaffolds in the form of hydrogels, polymer layers, and pre-formed macro-porous scaffolds (Fig. 6.1).

The scaffolds are an important component in the *in vitro* construction of the cardiac patch as they provide the physical support and biological cues and instruct tissue formation. Early studies in cardiac tissue engineering investigated scaffolds fabricated from natural polymers, such as alginate, chitosan, collagen, and gelatin, or from synthetic polyesters and polyethers. To improve survival and cell organization in these scaffolds, the cardiac cells were frequently seeded mixed with Matrigel to enable cell adhesion. However, Matrigel, being a gelatinous protein mixture secreted by mouse tumor

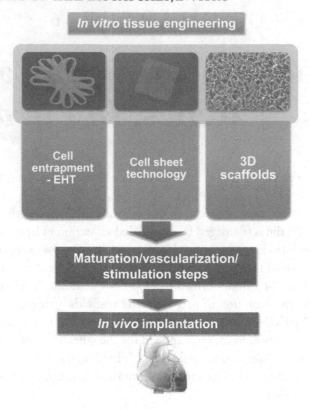

Figure 6.1: Strategies for reconstructing the cardiac patch. Three major strategies are used for preparation of cardiac patches: cell entrapment in hydrogels (EHT), cell sheet engineering, and cell seeding in preformed macroporous 3D scaffolds. After initial culture, various stimulation strategies could be applied (i.e., perfusion bioreactors, electrical stimulation) in order to improve cell organization and tissue maturation. Finally, the constructs can be implanted into infarcted heart. See text for more details.

cells, contains growth factors and unknown proteins that limit its desirability for experiments, as they require precise conditions or implantation into human beings. Consequently, scaffold research in recent years has advanced to designing, synthesizing, and modifying materials to selectively and spatially interact with cells through defined bio-molecular recognition events. More recently, with the advancements in nano- and micro-fabrication techniques, these tools have been employed in the fabrication of scaffolds bio-mimicking the physical structure and architecture of cardiac ECM. In particular, these technologies have been adapted for enabling and promoting of patch vascularization.

As a tribute to the pioneers in this exciting field, we first introduce their works which lay the foundation for cardiac tissue engineering.

6.2   CARDIAC CELL ENTRAPMENT IN HYDROGELS—ENGINEERED HEART TISSUE (EHT)

In this strategy, cardiac cells are mixed with a liquid form of the biomaterial, followed by its solidification to create a mixed 3D cell hydrogel. This strategy has been developed by Eschenhagen and collaborators, who cultured embryonic cardiac myocytes in a collagen type I matrix to produce a coherently contracting 3D Engineered Heart Tissue (EHT) (Fig. 6.2) [2].

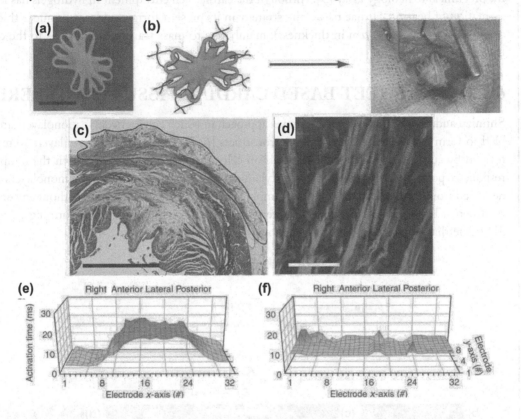

Figure 6.2: An engineered heart tissue (EHT) constructed by cell entrapment method. **(a)** Multiloop EHT ready for *in vivo* engraftment. **(b)** EHTs fixed on the recipient heart. **(c)** Four weeks after EHT engraftment, the construct formed compact and oriented heart muscle on the top of the infarct scar. **(d)** Laser confocal microscopy showing the highly differentiated sarcomeric organization of engrafted cardiomyocytes (actin, green; nuclei, blue). **(e-f)** Representative plots of epicardial activation times in sham-operated **(e)** and EHT-engrafted hearts **(f)**, showing undelayed coupling of EHT to the host myocardium. Reprinted with permission from [3].

Later, the same group improved the cardiac patch by suspending cardiomyocytes from neonatal rats in a mixture of collagen type I and Matrigel and casting the solidified matrix into circular molds [4]. The construct was then subjected to a mechanical stretch, stimulating the tissue to differentiate and mature. The result was a ring-shape EHT that displayed important hallmarks of differentiated myocardium, such as a striated shape and an electrical response to a β-agonist [4]. Implantation of the cardiac patch into infarcted hearts showed undelayed electrical coupling to the native myocardium without arrhythmias, prevented LV dilation, induced systolic wall thickening, and improved cardiac function [3]. The EHT construct has also been found to be a valuable tool for preclinical toxicology assay [5]. Although the cardiac cell entrapment in hydrogels has led to the formation of beating cardiac tissue, the strategy in its present form could not promote thick tissue assembly (i.e., over 100 μm in thickness), mainly due to mass transfer limitations to the center of the construct.

6.3   CELL SHEET-BASED CARDIAC TISSUE ENGINEERING

Shimizu and co-workers developed a novel approach in which cardiomyocyte monolayers are assembled to form multi-layered heart muscle constructs [6, 7]. The cardiomyocyte layers were initially formed by cultivating isolated cardiac cells on cell culture surfaces, grafted with the temperature-responsive polymer, poly ($N-$isopropylacrylamide) (PIPAAm). Confluent cell monolayers could be detached from the surface as a cell sheet simply by reducing the temperature, without any enzymatic treatments. The released cell layers (2D) were overlaid one on top of the other, forming a multi-layer 3D tissue that began to pulsate simultaneously (Fig. 6.3) [6].

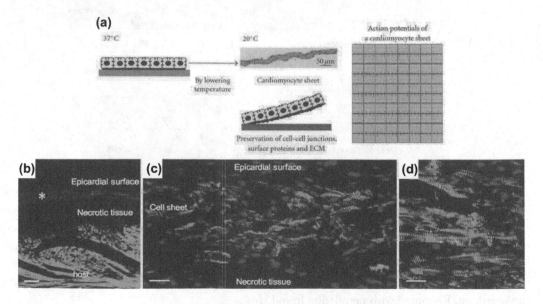

Figure 6.3: Reprinted with permission from [7, 8]. *Caption on the next page.*

Figure 6.3: Cell sheet-based cardiac tissue engineering. **(a)** Using temperature-responsive culture surfaces, cell structures are preserved and cultured cardiomyocytes are released as a contiguous cell sheet. A microphotograph shows the cross-sectional view of a cardiomyocyte sheet. Electrograms show the spontaneous action potentials of the cardiomyocyte sheet. **(b-d)** Cardiac tissue sheet grafts using 3D cell sheet manipulation technique integrate with the host heart. Laser confocal microscopic view of the grafted cell sheet and host heart, stained for α-actinin (green), anti-Cx-43 (red), and DAPI (blue) to stain the nucleus. **B.** An α-actinin-positive cell sheet can be seen at the epicardial surface of the necrotic tissue (*). **c.** Clear striation pattern of α-actinin and diffuse Cx-43 staining are seen in the cell sheet. **d.** The host image below the necrotic area. Scale bar: 20 μm. Reprinted with permission from [7, 8].

After implantation on the infarcted myocardium, gap junction formation and bi-directional smooth action potential propagation between the host heart and the grafted cardiac sheets were observed, without evidence of arrhythmias, suggesting functional and electrical graft integration (Fig. 6.3) [8, 9]. Cell sheet transplantation improved cell survival, angiogenesis and cardiac function, compared to dissociated cell injection [10]. Clinical trials using this technology have been initiated in other tissue applications (esophagus, cornea).

6.4   MYOCARDIAL TISSUE GRAFTS CREATED IN PREFORMED IMPLANTABLE SCAFFOLDS

In this approach, cardiac cells are seeded within pre-formed macro-porous scaffolds. The scaffolds are generally prepared by cross-linking (chemical or physical) of biomaterial solution into the desired shape, with a subsequent solidification step and/or drying/freeze-drying. This platform offers important advantages over cell entrapment during hydrogel fabrication, such as the ability to design and control the porosity of the scaffold prior to cell seeding and reduced exposure of the seeded cells to stress during mixing and molding of the scaffold. Furthermore, the pre-designed scaffolds can direct the seeded cells to organize into the gross conformation of native cardiac tissue, for example by influencing cardiac cell alignment in the porous structure.

Morphologically, the scaffold should be porous with an interconnecting pore structure to enable accommodation of a large number of cells and their subsequent organization into a functioning tissue [11]. Pore size should be at least 50 μm to allow high mass transport during *in vitro* culture and vascularization of the scaffold following implantation, so as to supply the seeded cells with nutrients and to remove secretions [12]. At the same time, the scaffold should present the desired mechanical properties and enable its handling in culture and during transplantation.

We were among the first groups to report on a successful implantation of cardiac cell-seeded porous alginate scaffolds into infarcted rat hearts [13, 14]. We found that the seeded fetal rat cardiac cells retained viability within the scaffolds and within 24 hr formed multicellular beating cell clusters [15]. Following implantation of the cellular constructs into the infarcted myocardium, some of the cells appeared to differentiate into mature myocardial fibers. The graft and surrounding area were populated with a large number of newly formed blood vessels, consequently leading to attenuation in LV dilatation and improved heart function (Fig. 6.4) [14].

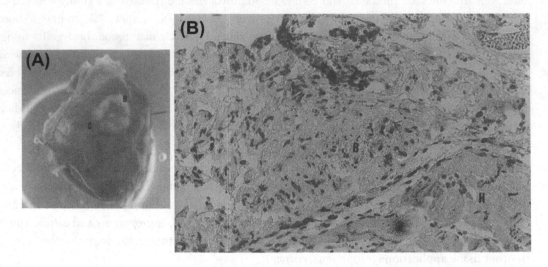

Figure 6.4: Bioengineered cardiac patch implantation on infarct. **A.** Photograph of heart at week nine after patch implantation. Visual examination revealed intense neovascularization into implanted patch (B, biograft). Note a coronary branch (C, coronary) that supplies and covers the patch with an extensive network of vessels. **B.** Microscopic image of integrated patch, immuno-stained for connexin-43 (Cx-43) (brown). Cx-43 was localized in normal parallel arrangements in host myocardium (H, host) and randomly oriented in the patch (B, biograft) (original magnification × 200). Reprinted with permission from [14].

The efficacy of this strategy was further proven with patches prepared from various other materials, natural (i.e., collagen, gelatin) or synthetic (i.e., poly(2-hydroxyethyl methacrylate-co-methacrylic acid), emphasizing the potential of this approach for myocardial repair [16, 17, 18, 19, 20].

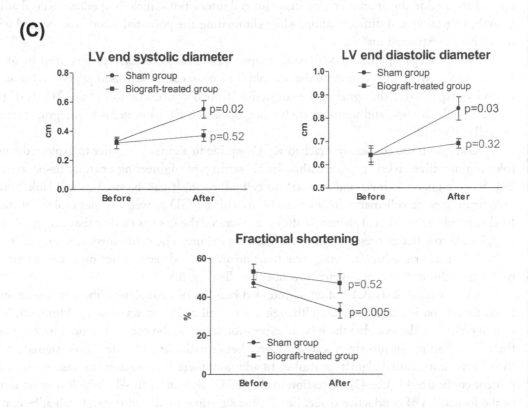

Figure 6.4: Bioengineered cardiac patch implantation on infarct. **C.** Results of echocardiography study. Fractional shortening = [(LV end diastolic diameter − LV end systolic diameter)/LV end diastolic diameter] × 100. P values for the differences between measurements before and after implantation. Reprinted with permission from [14].

6.5   BIOMIMETIC SCAFFOLDS AND INTEGRATION OF CELL-MATRIX INTERACTIONS

Along with the use of ECM-extracted biological materials as scaffolds, which intrinsically retain signals important for cell-matrix interactions, major efforts have been invested in the intelligent design of novel synthetic materials bio-inspired by the ECM interactions. These materials were created to provide the appropriate interface for cell interactions, including enhanced cell adhesion, growth, migration and differentiation, while eliminating the potential hazards associated with the use of ECM-extracted materials.

One strategy applied to achieve such appropriate material surfaces has been by attaching ECM-derived synthetic peptides to the scaffold. The most commonly used peptide is the sequence Arg-Gly-Asp (RGD), the signaling domain derived from fibronectin and laminin [21]. RGD peptide mediates cell adhesion and signaling via binding of various ECM proteins to integrin receptors on the cell surface [22].

Our group has covalently attached RGD peptide to alginate, in order to explore the possible role of immobilized RGD peptide within the 3D settings of engineering a cardiac tissue *in vitro* [23, 24]. Since alginate is inert and resistant to cell adhesion, it can be used as a "blank canvas" to investigate specific cell-matrix interactions by attaching RGD as well as other biological cues. The RGD peptide decoration of alginate scaffolds accelerated the *in vitro* cardiac tissue regeneration and contributed to a better preservation of the tissue in culture. The cardiomyocytes were able, within a short time after seeding, to reorganize their myofibrils and reconstruct myofibers composed of multiple cardiomyocytes in a typical myofiber bundle (Fig. 6.5).

Cardio-fibroblasts (CF) often surrounded bundles of cardiac myofibers in a manner similar to that of native cardiac tissue (although the overall cell mass was lower). Moreover, Western immunoblotting showed that the relative expression levels of the contractile protein α-actinin and the cell–cell adhesion protein N-cadherin were better maintained in the RGD–alginate scaffolds than in the unmodified alginate scaffolds. In addition, there was massive increase in gap junction protein connexin-43 (Cx-43) expression in the RGD–alginate scaffolds, which may be indicative of the formation of conductive tissue. Finally, the signaling mediated through cell adhesion to the RGD–alginate scaffold prevented cell apoptosis, further emphasizing the critical role of cell-matrix interactions in cell survival, organization, and tissue maturation [23].

More recently, we showed that the integration of an additional molecular signal provided by heparin-binding peptides (HBPs) promotes the formation of improved cardiac tissue *in vitro* [25]. HBPs, with the sequence XBBXBX and XBBBXXBX, where B is a basic amino acid and X is a hydropathic amino acid, have been shown to bind cell surfaces via the cellular syndecans, which bear 3-7 covalently attached heparin/heparan-sulfate chains [26]. Thus, we compared cardiac tissue engineering of neonatal rat cardiac cells cultivated within macro-porous alginate scaffolds bearing the two peptides, RGD and HBP, to those formed in a single peptide-attached or in unmodified alginate scaffolds [25]. Peptide modification did not affect scaffold architecture, pore size, and stiffness. Examination of cardiac cell morphology 14 days after seeding in the different scaffolds revealed the

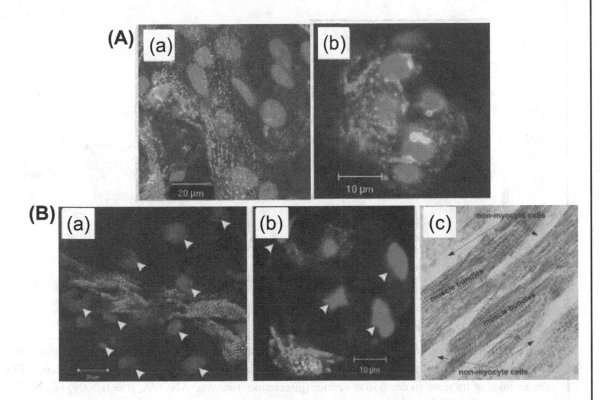

Figure 6.5: The effects of immobilized RGD peptide in alginate scaffolds on cardiac tissue engineering. **A.** Confocal microscopy images of cardiac cells cultivated in RGD-immobilized (a) and unmodified alginate scaffolds (b) after 12 days of cultivation, showing striking differences in cell organization. **B.** Relative locations of cardiomyocytes and CF in (a) RGD-immobilized and (b) unmodified alginate scaffold; (c) the native adult cardiac tissue stained for Troponin T (Tn-T). Arrowheads denote cell nuclei of CFs. Note the CFs surrounding cardiomyocyte bundles. Reprinted with permission from [23].

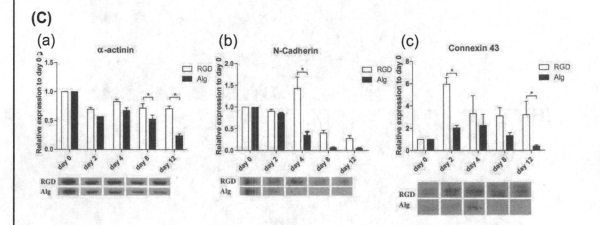

Figure 6.5: The effects of immobilized RGD peptide in alginate scaffolds on cardiac tissue engineering. **C.** Western blots and quantification of representative cardiomyocyte and nonmyocyte proteins. The relative folds of increase to day 0 in α-actinin (interaction, two-way ANOVA, $p = 0.0009$) (a), N-Cad ($p = 0.0001$) (b) and Cx-43 ($p = 0.0525$) (c) levels in the two constructs are shown. In staining: green – α-actinin, red – propidium iodide (PI). Reprinted with permission from [23].

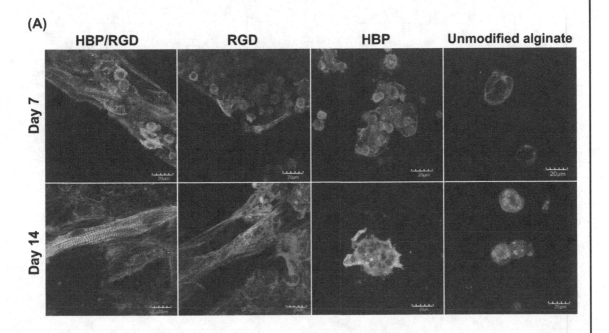

Figure 6.6: Integration of multiple cell-matrix interactions into alginate scaffolds promotes cardiac tissue organization. **A.** Cardiac cell organization, on days 7 and 14 post cell seeding. Cardiomyocytes are positive for α-actinin (green), while all cells are stained for F-actin (red) and nuclei (blue) (bar: 20 μm).

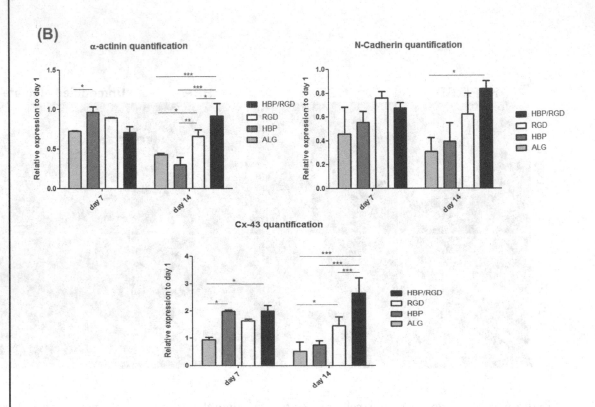

Figure 6.6: Integration of multiple cell-matrix interactions into alginate scaffolds promotes cardiac tissue organization. **B.** Protein expression of representative cardiac cell markers. Asterisks denote significant difference (by 2-way ANOVA), $^*p < 0.05$, $^{**}p < 0.01$ and $^{***}p < 0.005$ (Bonferroni's pos-hoc test was used for comparison between the groups). Reprinted with permission from [25].

characteristic striated fiber organization mainly in the RGD/HBP-attached cell constructs, while in HBP-attached and unmodified alginate cell constructs no such structures were observed (Fig. 6.6).

By histology, the 14-day constructs in HBP/RGD-attached scaffolds presented an isotropic arrangement of the fibers in the form of consistent tissue, while in the RGD-attached scaffold no such arrangement was seen. Western blot analysis of selected cardiac markers (sarcomeric α-actinin, N-cadherin, and Cx-43) showed preservation of these proteins and an increase in expression level of Cx-43 with time in the HBP/RGD-attached scaffolds, further supporting the notion of contractile muscle formation and tissue maturation (Fig. 6.6) [25]. Collectively, the results of these studies emphasize the importance of integration of multiple and complementary signals into 3D matrices for the recreation of the cardiac microenvironment and successful formation of an engineered cardiac tissue *in vitro*.

6.6   MICRO- AND NANOTECHNOLOGY FOR SCAFFOLD FABRICATION

Bulk materials design and relatively simple preparation schemes for scaffold fabrication have proven to be not only feasible, but also effective for promoting cardiac tissue engineering. However, growing functional tissue constructs still suffer from unmet needs and require a greater detail of control over the microenvironment wherein the cells reside. Micro- and nano-fabrication techniques, including micro- and nano-patterning of topographical or biochemical cues, offer advanced tools to attain such rigorously controlled cell environments and have significantly contributed to the engineering of functional cardiac tissues [27, 28].

In particular, nano- and micro-fabrication techniques have been applied to the fabrication of scaffolds with internal structure and architecture suitable for creating vascularized tissues. A functional cardiac patch with clinically relevant dimensions can be achieved if engineered with a proper vasculature to supply the cells with oxygen and nutrients as well as to instruct tissue formation. With micro-fabrication techniques in hand, patterns of vascular tree-like organization can be formed in scaffolds, subsequently seeded with endothelial cells to form a rudimentary vasculature. Such vascular networks should be continuous and may facilitate *in vivo* tissue integration through immediate anastomosis to the host vasculature [27]. Maidhof *et al* used developed channeled scaffolds made from the porous elastomer poly(glycerol sebacate) (PGS) [29]. A parallel array of channels with diameter of 250 μm (with 500 μm wall-to-wall spacing) was formed by computerized carbon dioxide laser piercing. The group created a cardiac patch with parallel channels lined with endothelial cells using two-stage seeding technique. At first, by stacking the scaffolds on top of one another in perfusion cartridges, thus effectively closing off the channels during perfusion seeding, a uniform spatial distribution of rat cardiomyocytes in the scaffold was achieved. At second stage, lining of the channels in the scaffold with rat aortic endothelial cells was attained by perfusion seeding of single construct. Combination of these perfusion seeding techniques allows homogeneous seeding of cardiomyocytes, which remain shielded in the scaffold pores from direct hydrodynamic shear, and results in the preferential attachment of ECs to channel wall. Collectively, creation of cardiac patches with such primitive vasculature allows engineering of thicker cardiac constructs *in vitro*, enhancing the ability of the construct to survive and function upon *in vivo* implantation [29].

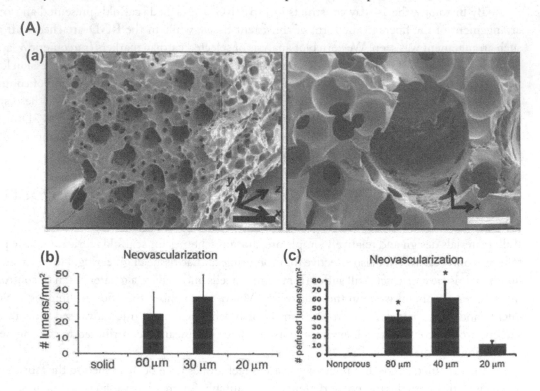

Figure 6.7: Implementation of micro/nano-technologies in cardiac tissue engineering. **A.** Bimodal proangiogenic scaffolds. **a.** SEM images of bimodal pHEMA-co-MAA scaffolds. Final scaffold design consists of 60-μm channels spaced 60 μm apart. Channel walls contain spherical pores with a 30-μm diameter and 15-μm interconnects. **b.** Quantification of RECA-1 (rat endothelial cell antigen-1)-positive lumen structures. **c.** Density of functional vessels (perfused by biotinylated lectin) was quantified over the entire implant with 40- and 80-μm porous scaffolds having significantly higher densities than nonporous and 20-μm porous scaffolds [16]. Reprinted with permission from [30, 31].

Madden *et al* fabricated parallel channeled scaffolds from poly(2-hydroxyethyl methacrylate-co-methacrylic acid) (pHEMA-co-MAA) by micro-templating. These bimodal scaffolds consisted of interconnected spherical pore regions (for angiogenesis) and small-diameter channels (Fig. 6.7A) [16]. Channel size and spacing were controlled by varying the dimensions of the template (45- to 150-μm diameter). The smallest diameter compatible with reliable cell seeding, 60 μm, was chosen, to alleviate mass transfer issues within the channel. Channel spacing of 60 μm was selected to allow introduction of pores ranging from 20–40 μm. In this configuration, channels account for 25% of scaffold volume; the remaining 75% is a porous network of ∼60% void space. Channel

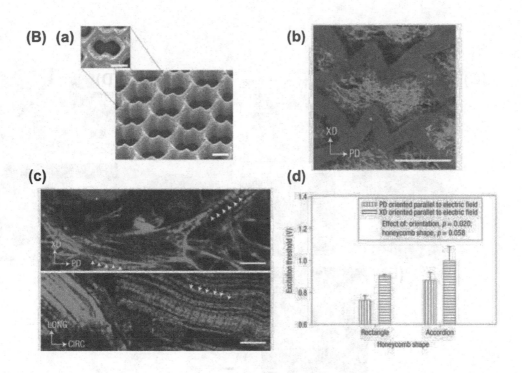

Figure 6.7: Implementation of micro/nano-technologies in cardiac tissue engineering. **B.** Accordion-like honeycomb scaffolds for tissue engineering of cardiac anisotropy. **a.** Scanning electron micrographs show the structure of excimer laser micro-ablated accordion-like PGS scaffolds. **b.** Low-magnification image of a representative graft demonstrated pores completely filled by neonatal rat neonatal heart cells grossly aligned in parallel to the preferred direction (one week cultivation). **c.** Higher-magnification images of neonatal rat heart cells cultured on scaffolds demonstrating the presence of some elongated neonatal rat heart cells and cross-striations (white arrows) similar but significantly less developed than those present in control specimens (bottom) of adult rat right ventricular myocardium. **d.** Electrical field stimulation demonstrated directionally dependent excitation thresholds for grafts based on anisotropic scaffolds. The excitation threshold was significantly lower ($p =0.02$) when the scaffold preferred direction was oriented parallel to the electric field, with no consistent differences observed in isotropic square honeycombs (not shown). Reprinted with permission from [30, 31].

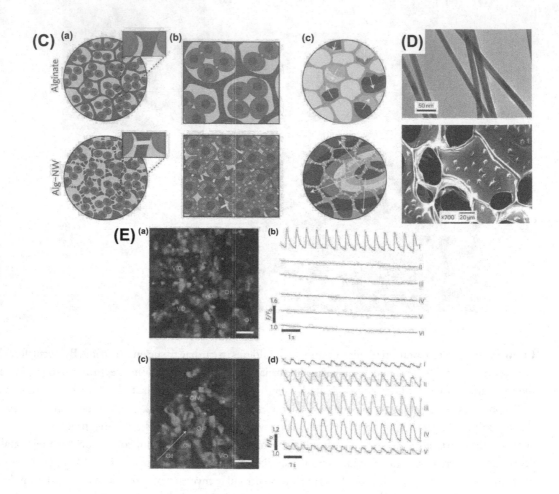

Figure 6.7: *Caption on the next page.* Reprinted with permission from [30, 31].

Figure 6.7: Implementation of micro/nano-technologies in cardiac tissue engineering. **C.** Schematic overview of three-dimensional engineered nanowired (NW) cardiac tissue. **a).** Isolated cardiomyocytes are cultured in either pristine alginate or Alg–NW composites. Insets highlight the components of the engineered tissue: cardiac cells (red), alginate pore walls (blue), and gold nanowires (yellow). **b).** Whereas cardiomyocytes in pristine alginate scaffolds (top) typically form only small clusters that beat asynchronously and with random polarization, Alg–NW scaffolds (bottom) can exhibit synchronization across scaffold walls, throughout the entire scaffold. **c).** Cardiomyocytes cultured in alginate scaffolds (top) form small beating clusters, but synchronously beating cardiomyocytes in Alg–NW composites (bottom) have the potential to form organized cardiac-like tissue. Colors, contour lines, and arrows represent the spatial and temporal evolution of the signal maximum. **D.** Incorporation of nanowires within alginate scaffolds. Transmission electron microscopy images of a typical distribution of gold nanowires, which exhibited an average length of $\sim$1 μm and average diameter of 30 nm. SEM revealed that the nanowires (1 mg ml^{-1}) assembled within the pore walls of the scaffold into star-shaped structures with a total length scale of 5 μm. The assembled wires were distributed homogeneously within the matrix at a distance of $\sim$5 μm from one another (bottom). **E.** Calcium transient propagation within engineered tissues. Calcium transient was assessed at specified points (white circles) by monitoring calcium dye fluorescence (green). **a).** Sites monitored in pristine scaffold, where site I is the stimulation point. **b).** Calcium transients were only observed at the stimulation point in the unmodified scaffold. F/F_0 refers to measured fluorescence normalized to background fluorescence. **c).** Sites monitored in an Alg–NW scaffold. The stimulation point was 2 mm diagonally to the lower left of point I (that is, off the figure). The white arrow represents the direction of propagation. **d).** Calcium transients were observed at all points. Reprinted with permission from [30, 31].

domains are set to promote bundled orientation of cardiomyocytes, and spherical pore domains for mass transfer and invading vasculature. These bimodal scaffolds were seeded with human ES cell-derived cardiomyocytes and cultured *in vitro*. Cardiomyocytes survived and proliferated for two weeks in scaffolds, reaching adult heart densities. Cardiac implantation of acellular scaffolds with pore diameters of 30–40 μm showed angiogenesis and reduced fibrotic response, coinciding with a shift in macrophage phenotype toward the prohealing M2 state, suggesting that this shift could be responsible for proangiogenic response elicited by scaffold material [16]. This work establishes a foundation for spatially controlled cardiace tissue engineering by providing discrete micro-compartments for cardiomyocytes and stroma in a scaffold that enhances vascularization and integration while controlling the inflammatory response.

Nano- and micro-fabrication techniques were also implemented to develop anisotropic cardiac patches. The ventricular myocardium is a richly vascularized, quasi-lamellar tissue in which functional syncytia of cardiomyocytes (organized in cardiac muscle fibers) are interwoven within collagen. Hierarchically, the cardiac muscle fibers are surrounded and coupled by endomysial collagen sheaths that are bundled within a honeycomb-like network of undulated perimysial collagen fibers. These features yield directionally dependent electrical and mechanical properties collectively termed cardiac anisotropy [32]. The bulk-designed porous scaffolds described so far could not support the development of a substantial anisotropic cardiac tissue patch since their interior structure is isotropic in nature. To achieve anisotropic cardiac tissue constructs, aligned PLGA scaffolds were developed by leaching of aligned sucrose templates (incorporated into the PLGA solution during fabrication) and then the scaffolds were coated with fibronectin to allow cell attachment. The aligned scaffolds were seeded with neonatal rat cardiac cells and cultured in rotating high-aspect-ratio vessel bioreactors (HARVs) for 6-14 days. With time, the cardiac cells aligned and interconnected inside the scaffolds and when stimulated by a point electrode, supported macroscopically continuous, anisotropic impulse propagation. By culture day 14, the ratio of conduction velocities along vs. across cardiac fibers reached a value of 2, similar to that in native neonatal ventricles, while the action potential duration and maximum capture rate, respectively, decreased to 120 ms and increased to $\sim$5 Hz. In contrast, randomly oriented cells in control cardiac monolayers supported only isotropic propagation, and these cultures deteriorated by day 9 [33]. Engelmayr *et al* used excimer laser microablation to create an accordion-like honeycomb microstructure in PGS, yielding porous, elastomeric scaffolds with controllable stiffness and anisotropy (Fig. 6.7B) [30]. The authors reasoned that accordion-like honeycomb scaffolds exhibiting distinct preferred and orthogonal, cross-preferred material directions could potentially: (1) match the anisotropic in-plane mechanical response of native myocardium within the physiologic regime, (2) provide low in-plane resistance to contraction and (3) provide an inherent structural capacity to guide cardiomyocyte orientation in the absence of external stimuli. These scaffolds with cultured neonatal rat heart cells demonstrated closely matched mechanical properties compared to native adult rat right ventricular myocardium, with stiffnesses controlled by polymer curing time. Heart cell contractility inducible by electric field stimulation showed directionally dependent electrical excitation thresholds and greater heart cell alignment, compared to isotropic control scaffolds. Prototype bilaminar scaffolds with 3D interconnected pore networks yielded electrically excitable grafts with multi-layered neonatal rat heart cells [30].

Other than 3D microfabrication, geometrically controlled scaffolds can also be generated through other nanofabrication methods such as electrospinning, where biomaterials are spun into nanofibers to create a 3D scaffold closely mimicking the ECM network *in vivo*. The advantage of using nanofibers is that the created nanoscale mesh, although randomized, resembles more closely the native ECM than scaffolds created through microfabrication [27]. To enhance tissue organization, nanofibers can be organized unidirectionally to promote cell alignment. Orlova *et al* produced aligned nanofibres with prismatic cross sections (700–1000 nm in width and 300–500 nm in height) from polymethylglutarimide (PMGI) with electrospinning [34]. The alignment of the nanofibers was controlled by the collector which included a rectangular hole. The orientation and dimension of the rectangular holes controlled the orientation and length of the nanofibers. The positioning density of the nanofibers was controlled by varying the time of fiber deposition. Seeded cardiac cells proliferated and aligned into a contractile tissue. The elongation and alignment of the cardiac cells were characterized by the orientation of the α-actin filaments [34]. The significance of this work is the demonstration of the three-dimensionality of the scaffold and the resulting tissue as well as the structural anisotropy.

Nanostructures can also be used to induce stimulation and/or provide specific signals through the engineered tissue constructs. For example, one of the limitations of biomaterial-based porous matrices is that their pore walls limit cell-cell interaction and delay electrical signal propagation. As a result, the poor conductivity of these scaffolds limits the ability of the engineered cardiac patches to contract strongly as a unit, thus reducing patch functionality. You *et al* prepared hybrid hydrogel scaffolds made of a combination of thiol-2-hydroxyethyl metacrylate (thiol-HEMA) and HEMA. The scaffolds were made electrically conductive by impregnation of Au nanoparticles that were homogenously synthesized throughout a polymer template gel. The resulting conductive gels had Young's moduli more similar to myocardium relative to polyaniline and polypyrrole (generally used conductive polymers, that fail to mimic physiological mechanical properties), by 1-4 orders of magnitude. Neonatal rat cardiomyocytes exhibited two-fold increased expression of Cx-43 when grown on hybrid scaffolds relative to non-conductive HEMA scaffolds, with or without electrical stimulation [35]. In another attempt to confer electrical conductivity, Dvir *et al* created a nanocomposite alginate scaffold containing gold nanowires (with average length of ∼1 μm, to cross and bridge the electrically resistant scaffold pore walls) (Fig. 6.7, C-E). The nanocomposites exhibited improved mechanical properties because the nanomaterials, which interact with the polymer matrix, may act as reinforcements. Tissues of cardiac cells (rat neonatal cardiomyocytes and fibroblasts) grown on these composite matrices were thicker and better aligned than those grown on pristine alginate. When electrically stimulated, the cells in these tissues contracted synchronously, as evaluated by calcium imaging showing calcium transient recordings at various sites in the nanocomposite scaffolds, as opposed to calcium transients revealed at the stimulation site only in pristine alginate scaffolds. Furthermore, higher levels of the proteins involved in muscle contraction and electrical coupling (troponin-I and Cx-43) were detected in the composite matrices [31].

Our group recently explored alginate nanocomposite scaffolds incorporating magnetic nanoparticles (MNP) to provide means of physical stimulation to endothelial cells (see Chapter 7).

6.7   SUMMARY AND CONCLUSIONS

The goal of this chapter was to introduce the *in vitro* strategies of cardiac tissue engineering to create cardiac patches with thickness appropriate to replace a whole scar. Such cardiac patches have not yet been achieved, but the advances made in recent years indicate that we are on the right track to achieve this goal. Herein, we mainly focused on describing the engineered cardiac tissue as different aspects of scaffold fabrication were optimized, from the synthesis and integration of cell-matrix interactions to the nano/micro-fabrication of channels in the scaffold for enabling its vascularization. Clearly, the application of nano/micro-technology and science to scaffold fabrication is a new frontier in cardiac tissue engineering research. We expect that using these new technologies, the scaffolds created would better mimic the cardiac ECM leading to the development of thick functional cardiac tissue.

BIBLIOGRAPHY

[1] Zimmermann WH, Didie M, Doker S, Melnychenko I, Naito H, Rogge C, et al. Heart muscle engineering: an update on cardiac muscle replacement therapy. Cardiovasc Res. 2006;71:419–29. DOI: 10.1016/j.cardiores.2006.03.023 63

[2] Eschenhagen T, Fink C, Remmers U, Scholz H, Wattchow J, Weil J, et al. Three-dimensional reconstitution of embryonic cardiomyocytes in a collagen matrix: a new heart muscle model system. Faseb J. 1997;11:683–94. 65

[3] Zimmermann WH, Melnychenko I, Wasmeier G, Didie M, Naito H, Nixdorff U, et al. Engineered heart tissue grafts improve systolic and diastolic function in infarcted rat hearts. Nature medicine. 2006;12:452–8. DOI: 10.1038/nm1394 65, 66

[4] Zimmermann WH, Schneiderbanger K, Schubert P, Didie M, Munzel F, Heubach JF, et al. Tissue engineering of a differentiated cardiac muscle construct. Circ Res. 2002;90:223–30. DOI: 10.1161/hh0202.103644 66

[5] Schaaf S, Shibamiya A, Mewe M, Eder A, Stohr A, Hirt MN, et al. Human engineered heart tissue as a versatile tool in basic research and preclinical toxicology. PLoS One. 2011;6:e26397. DOI: 10.1371/journal.pone.0026397 66

[6] Shimizu T, Yamato M, Isoi Y, Akutsu T, Setomaru T, Abe K, et al. Fabrication of pulsatile cardiac tissue grafts using a novel 3-dimensional cell sheet manipulation technique and temperature-responsive cell culture surfaces. Circ Res. 2002;90:e40. DOI: 10.1161/hh0302.105722 66

[7] Haraguchi Y, Shimizu T, Yamato M, Okano T. Regenerative therapies using cell sheet-based tissue engineering for cardiac disease. Cardiol Res Pract. 2011;2011:845170. DOI: 10.4061/2011/845170 66, 67

[8] Furuta A, Miyoshi S, Itabashi Y, Shimizu T, Kira S, Hayakawa K, et al. Pulsatile Cardiac Tissue Grafts Using a Novel Three-Dimensional Cell Sheet Manipulation Technique Functionally Integrates With the Host Heart, In Vivo. Circulation Research. 2006;98:705–12. DOI: 10.1161/01.RES.0000209515.59115.70 66, 67

[9] Sekine H, Shimizu T, Kosaka S, Kobayashi E, Okano T. Cardiomyocyte bridging between hearts and bioengineered myocardial tissues with mesenchymal transition of mesothelial cells. J Heart Lung Transplant. 2006;25:324–32. DOI: 10.1016/j.healun.2005.09.017 67

[10] Sekine H, Shimizu T, Dobashi I, Matsuura K, Hagiwara N, Takahashi M, et al. Cardiac Cell Sheet Transplantation Improves Damaged Heart Function via Superior Cell Survival in Comparison with Dissociated Cell Injection. Tissue Eng Part A. 2011. DOI: 10.1089/ten.tea.2010.0659 67

[11] Shapiro L, Cohen S. Novel alginate sponges for cell culture and transplantation. Biomaterials. 1997;18:583–90. DOI: 10.1016/S0142-9612(96)00181-0 67

[12] Zmora S, Glicklis R, Cohen S. Tailoring the pore architecture in 3-D alginate scaffolds by controlling the freezing regime during fabrication. Biomaterials. 2002;23:4087–94. DOI: 10.1016/S0142-9612(02)00146-1 67

[13] Cohen S, Leor J. Rebuilding broken hearts. Biologists and engineers working together in the fledgling field of tissue engineering are within reach of one of their greatest goals: constructing a living human heart patch. Scientific American. 2004;291:44–51. 67

[14] Leor J, Aboulafia-Etzion S, Dar A, Shapiro L, Barbash IM, Battler A, et al. Bioengineered cardiac grafts. A new approach to repair the infarcted myocardium? Circulation. 2000;102 (supplII):56–61. 67, 68, 69

[15] Dar A, Shachar M, Leor J, Cohen S. Optimization of cardiac cell seeding and distribution in 3D porous alginate scaffolds. Biotech and Bioeng. 2002;80:305–12. DOI: 10.1002/bit.10372 67

[16] Madden LR, Mortisen DJ, Sussman EM, Dupras SK, Fugate JA, Cuy JL, et al. Proangiogenic scaffolds as functional templates for cardiac tissue engineering. Proceedings of the National Academy of Sciences of the United States of America. 2010;107:15211–6. DOI: 10.1073/pnas.1006442107 68, 76, 79

[17] Radisic M, Park H, Shing H, Consi T, Schoen FJ, Langer R, et al. Functional assembly of engineered myocardium by electrical stimulation of cardiac myocytes cultured on scaffolds. Proceedings of the National Academy of Sciences of the United States of America. 2004;101:18129–34. DOI: 10.1073/pnas.0407817101 68

[18] Kutschka I, Chen IY, Kofidis T, Arai T, von Degenfeld G, Sheikh AY, et al. Collagen matrices enhance survival of transplanted cardiomyoblasts and contribute to functional improvement of ischemic rat hearts. Circulation. 2006;114:I167–73. DOI: 10.1161/CIRCULATIONAHA.105.001297 68

[19] Li RK, Jia ZQ, Weisel RD, Mickle DA, Choi A, Yau TM. Survival and function of bioengineered cardiac grafts. Circulation. 1999;100:II63–9. DOI: 10.1161/01.CIR.100.suppl_2.II-63 68

[20] Ye KY, Black LD, 3rd. Strategies for Tissue Engineering Cardiac Constructs to Affect Functional Repair Following Myocardial Infarction. J Cardiovasc Transl Res. 2011. DOI: 10.1007/s12265-011-9303-1 68

[21] Shin H, Jo S, Mikos AG. Biomimetic materials for tissue engineering. Biomaterials. 2003;24:4353–64. DOI: 10.1016/j.addr.2007.08.041 70

[22] Rosso F, Giordano A, Barbarisi M, Barbarisi A. From cell-ECM interactions to tissue engineering. J Cell Physiol. 2004;199:174–80. DOI: 10.1002/jcp.10471 70

[23] Shachar M, Tsur-Gang O, Dvir T, Leor J, Cohen S. The effect of immobilized RGD peptide in alginate scaffolds on cardiac tissue engineering. Acta Biomater. 2011;7:152–62. DOI: 10.1016/j.actbio.2010.07.034 70, 71, 72

[24] Tsur-Gang O, Ruvinov E, Landa N, Holbova R, Feinberg MS, Leor J, et al. The effects of peptide-based modification of alginate on left ventricular remodeling and function after myocardial infarction. Biomaterials. 2009;30:189–95. DOI: 10.1016/j.biomaterials.2008.09.018 70

[25] Sapir Y, Kryukov O, Cohen S. Integration of multiple cell-matrix interactions into alginate scaffolds for promoting cardiac tissue regeneration. Biomaterials. 2011;32:1838–47. DOI: 10.1016/j.biomaterials.2010.11.008 70, 74, 75

[26] Cardin AD, Weintraub HJ. Molecular modeling of protein-glycosaminoglycan interactions. Arteriosclerosis. 1989;9:21–32. DOI: 10.1161/01.ATV.9.1.21 70

[27] Zhang B, Xiao Y, Hsieh A, Thavandiran N, Radisic M. Micro- and nanotechnology in cardiovascular tissue engineering. Nanotechnology. 2011;22:494003. DOI: 10.1088/0957-4484/22/49/494003 75, 80

[28] Dvir T, Timko BP, Kohane DS, Langer R. Nanotechnological strategies for engineering complex tissues. Nature nanotechnology. 2011;6:13–22. DOI: 10.1038/nnano.2010.246 75

[29] Maidhof R, Marsano A, Lee EJ, Vunjak-Novakovic G. Perfusion seeding of channeled elastomeric scaffolds with myocytes and endothelial cells for cardiac tissue engineering. Biotechnol Prog. 2010;26:565–72. DOI: 10.1002/btpr.337 75, 76

[30] Engelmayr GC, Jr., Cheng M, Bettinger CJ, Borenstein JT, Langer R, Freed LE. Accordion-like honeycombs for tissue engineering of cardiac anisotropy. Nat Mater. 2008;7:1003–10. DOI: 10.1038/nmat2316 76, 77, 78, 79, 80

[31] Dvir T, Timko BP, Brigham MD, Naik SR, Karajanagi SS, Levy O, et al. Nanowired three-dimensional cardiac patches. Nature nanotechnology. 2011;6:720–5. DOI: 10.1038/nnano.2011.160 76, 77, 78, 79, 81

[32] Holmes JW, Borg TK, Covell JW. Structure and mechanics of healing myocardial infarcts. Annu Rev Biomed Eng. 2005;7:223–53. DOI: 10.1146/annurev.bioeng.7.060804.100453 80

[33] Bursac N, Loo Y, Leong K, Tung L. Novel anisotropic engineered cardiac tissues: studies of electrical propagation. Biochemical and biophysical research communications. 2007;361:847–53. DOI: 10.1016/j.bbrc.2007.07.138 80

[34] Orlova Y, Magome N, Liu L, Chen Y, Agladze K. Electrospun nanofibers as a tool for architecture control in engineered cardiac tissue. Biomaterials. 2011;32:5615–24. DOI: 10.1016/j.biomaterials.2011.04.042 80

[35] You JO, Rafat M, Ye GJ, Auguste DT. Nanoengineering the heart: conductive scaffolds enhance connexin 43 expression. Nano letters. 2011;11:3643–8. DOI: 10.1021/nl201514a 81

CHAPTER 7

Perfusion Bioreactors and Stimulation Patterns in Cardiac Tissue Engineering

CHAPTER SUMMARY

Regeneration of a thick functional cardiac patch *in vitro* presents a major engineering challenge, i.e., the need for reconstruction of a dynamic 3D cellular microenvironment with the appropriate chemical and mechanical signals to induce cell differentiation, maturation, and assembly into a functional tissue. In this chapter, we describe the creative design of various dynamic cell microenvironments, which promote the development of a thick cardiac patch, ready to face the harsh conditions in an infarcted heart. Among these microenvironments are unique perfusion bioreactor systems that increase mass transfer through the developing cardiac tissue at the *in vitro* engineering stage, and the use of modules which provide mechanical and electrical stimulation and induce the formation of a thick contractile cardiac tissue, *in vitro*.

7.1 INTRODUCTION

Engineering of a full thickness functional human cardiac muscle (~1-cm thick) is a great challenge for the tissue engineer. The cardiac muscle tissue constitutes densely packed cells and is composed of oxygen- and shear-sensitive cardiomyocytes. It is almost impossible to grow such thick and densely packed tissues under a static cultivation mode, relying only on a molecular diffusion for nourishing the cells. Anoxic conditions developed in the a-vascular cardiac cell constructs lead to cell death and the formation of thin external tissue layer at the periphery of the construct. In addition, accumulated data over the last years indicated that different physical cues, such as electrical signaling, mechanical stimulation (e.g., stretching of the 3D cellular constructs), are required to attain more functional cardiac tissues [1, 2, 3, 4]. These signals promote cardiomyocyte hypertrophy, increase the contractile protein content in tissue, and encourage the alignment of cells into myofibrillar structures, with contractility properties resembling those of a native cardiac muscle tissue.

The employment of bioreactors in tissue engineering has greatly advanced the field and contributed to the development of thicker and more functional cardiac tissue. Bioreactors are generally defined as devices in which biological and/or biochemical processes develop under closely monitored and tightly controlled environmental and operating conditions (e.g., pH, temperature, pressure, nutrient supply, and waste removal).

7.2   BIOREACTOR CULTIVATION OF ENGINEERED CARDIAC TISSUE

7.2.1   MASS TRANSFER IN 3D CULTURES

In vivo, oxygen and nutrients transport from the blood to the tissues is a diffusive process driven by a gradient of tension. Blood is delivered to every part of the body via vascular networks that serve as transport channels for convective flow and make it possible for diffusive nutrients to transfer across the cellular space and ECM more efficiently. When tissue engineering is performed outside the body *(ex vivo)* under static cultivation conditions, two main problems arise. First, the existence of a boundary layer around the cell construct decreases nutrients and oxygen diffusion to the construct surface. Second, the diffusion coefficient that is influenced by construct porosity and tortuosity is reduced to an effectiveness coefficient, as described in the following equation:

$$D_{eff} = D_N \frac{\varepsilon_p}{\tau}$$

Where D_{eff} is the effectiveness diffusion coefficient, D_N is the nutrient diffusion coefficient and ε_p and τ are the cell construct porosity and tortuosity, respectively. Taken together, the diffusion limitations cause serious problems in engineering a functional thick tissue. Cell constructs thicker than the diffusion distance suffer from poor tissue survival in the construct core due to anoxic conditions and lack of efficient waste removal.

Adding to this challenge is the relatively high cell density in the tissue engineered cardiac constructs. The consumption rate of nutrients and especially of dissolved oxygen in such a densely packed construct may be greater than molecular diffusion rate, leading to nutrient deprivation in the construct and cell death. Grodzinsky *et al* formulated a combined equation that generally follows Michaelis-Menten type dependence to describe nutrient concentration distribution L(x) in a tissue, while taking into account the transport rates across a tissue and the nutrient consumption rates by cells [5].

$$\frac{\partial L}{\partial t} = D \frac{\partial^2 L}{\partial x^2} - \frac{\rho k L}{K_m + L}$$

Where D is the diffusion coefficient, k is the maximal uptake rate constant per cell, K_m is the saturation constant, ρ is cell density, and x is the distance from the nutrient source.

Bio-inspired by nature, the transport of dissolved oxygen and nutrients can be improved by introducing medium convection to the cultivation system. The resulting nutrient concentration distribution is described then by: convection, diffusion, and consumption:

$$\frac{\partial L}{\partial t} = V_x \frac{\partial L}{\partial x} + D \frac{\partial^2 L}{\partial x^2} - \frac{\rho k L}{K_m + L}$$

Where V_x is the velocity of the fluid forced across the construct and transferring the essential molecules along the X-axis. Adding convection to the equation should improve nutrient and dissolved oxygen transport in the entire cell construct in a similar manner to the action of the capillary network in our body.

7.2.2   BIOREACTOR AS A SOLUTION FOR MASS TRANSFER CHALLENGE

The first-generation bioreactors for cell construct cultivation were adapted from vessels used to cultivate large volumes of concentrated cell suspensions, such as spinner flasks or rotating vessels. The bioreactors represented different patterns of vessel geometry, scaffold placing, and fluid dynamics to promote efficient fluid mixing and were successful in reducing the boundary layer surrounding the cell construct surface. In these systems, however, molecular diffusion was the main factor influencing the transport of dissolved oxygen and nutrients from the culture medium to the seeded cells in construct. Since oxygen diffusion alone can support only four to seven cell layers, cultivation of cell constructs under these condition resulted in only a 100 μm thick layer of viable and compact tissue (compatible with the oxygen diffusion distance) [6, 7, 8].

7.2.3   PERFUSION BIOREACTORS

In order to increase the thickness of viable cardiac tissue above ~100 μm, diffusional oxygen limitations have to be overcome and this has been accomplished by the use of perfusion bioreactors. The application of perfusion, by forcing the medium into the entire cell-seeded scaffolds, enables efficient mass transfer of oxygen and other soluble factors throughout the entire developing tissue, similar to the role of the vasculature in tissues.

Perfusion bioreactors operation is based on culture medium convection across a porous cell construct, thus providing a controllable, mechanically active environment for all seeded cells. It was shown that cultivation of engineered tissues in perfusion bioreactors improved the construct size, cellularity, and molecular composition. The bioreactors also made it possible to study effects of (i) mass transport of growth factors and nutrients, (ii) hydrodynamic load conditions, and (iii) physical stimulation, such as dynamic compression or cyclic stretch, on the developing tissue [9].

The group led by Vunjak-Novakovic was the first to conceive and apply interstitial medium flow in conjunction with porous scaffolds to improve the cardiac patch features. Their initial perfusion bioreactors were commercially available cartridges of 13 mm filter holders, connected to a peristaltic pump. A stainless steel screen was placed at the cartridge inlet to disperse the medium flow over the construct surface, while a nylon mesh was used to fix the downstream side of the construct [10]. Introduction of the pulsatile flow feature in such system improved the contractile properties and enhanced metabolic activity of the cells in the constructs, yet these features were confined mainly to the construct periphery [4, 11]. This may be explained by their use of mesh holders with 60-70% porous structure, resulting in a non-homogeneous environment within the system.

Our group designed a perfusion bioreactor system able to create homogenous medium flow in cell constructs by installing two cell construct-fixing nets, with an open area of 95.8% [12] (Fig. 7.1).

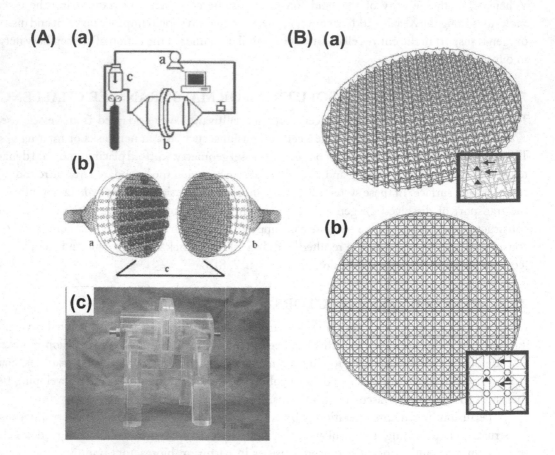

Figure 7.1: Perfusion bioreactor system. **A-a.** Overall view: computerized peristaltic pump (a) circulates the medium into the bioreactor vessel (b). The medium then proceeds to the reservoir (c), where it is oxygenated and heated to 37°C. **A-b.** Schematic representation of the bioreactor vessel. Two identical halves of the bioreactor (a, b) function as the inlet and the outlet of the cell construct compartment (c). Multiple cell constructs (blue) are fixed between two open pore nets. The angles of both halves were designed to be 60° to prevent the breakdown of the incoming flow. **A-c.** Picture of the bioreactor. **B.** Architecture of the nets holding the constructs in the bioreactor. Side (**a**) and upper view (**b**). The net architecture is an assembly of equally spaced, 380-square-based pyramidal structures (2 mm length × 2 mm height, each) with their heads pointing toward and holding the scaffolds. Four round openings (diameter =1 mm) are located at the corners of the square pyramid. The medium is perfused through these small holes toward the pyramid head at an angle of 60°. Arrows point to pyramidal heads and arrowheads to the pores at four corners of the base.

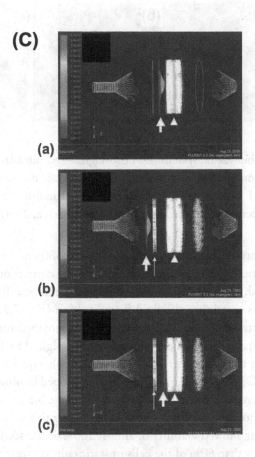

Figure 7.1: Perfusion bioreactor system. **C.** Profile of the fluid flow velocity in the perfusion bioreactor, constructed via the computerized fluid dynamics software Fluent after solving Navier-Stock equations. The velocity profiles in a perfusion bioreactor with no fluid-distributing mesh (**a**) or when approaching it in a bioreactor with the flow-distributing mesh (**b**) reveal a well-developed flow with large-velocity vectors (green) at the center of the bioreactor and smaller ones on the sides (blue). Once past the mesh (**c**), the flow is interrupted and transforms to undeveloped flow, with equal velocity vectors (blue line) along the bioreactor cross-section. Arrowhead indicates the location of the cellular constructs; thin, long arrows indicate the location of the flow-distributing mesh 1.5 cm upstream from the construct compartment, and thick, short arrows indicate the flow velocity profile. Velocity scale-bar is presented on the left. Dark blue represents the smallest vectors.

(D)

Figure 7.1: Perfusion bioreactor system. **D.** Histological and ultrastructural morphology of the engineered tissue in perfusion bioreactor. (**a**) H&E staining reveals massive striation and elongation of the cultured cells. (**b**) TEM images show defined Z-lines and multiple high-ordered sarcomeres and (**c**) intercalated disks between adjacent cardiomyocytes. Scale bars: a, 20 μm; b, 1 μm; c, 0.5 μm [12, 13].

The net architecture is an array of equally spaced, 380-square-based micro-pyramidal structures, able to hold multiple scaffolds simultaneously. At the corners of the square pyramid four round openings are positioned, enabling medium convection from the small holes toward the pyramid head, at an angle of 60°, thus sustaining the fluid flow direction (Fig. 7.1B). Using this net architecture, 99.88% of the cell construct volume is perfused by the culture medium. Our perfusion bioreactor was upgraded in another aspect, by insertion of micromesh, designed to disrupt the media flow for a split second before entering the cell constructs, in order to apply equal shear stress along the bioreactor cross-section (Fig. 7.1C). The mesh transforms the developed laminar fluid flow velocity profile into an undeveloped one, having equal velocity vectors along the bioreactor cross-section and subjecting the cellular constructs to an equal shear stress. When using this innovative perfusion bioreactor, the cardiac tissue constructs showed viability of almost 100% of the seeded cells after a seven-day cultivation period, while less than 60% of the cells in static cultures were viable at the end of this period. Furthermore, medium reaching the construct core by perfusion enabled the formation of homogenous, thick (>500 μm), cardiac tissue. Tissue architecture analysis by histochemistry revealed the presence of elongated and aligned cells with massive striation. Ultrastructural morphology analyses presented organized sarcomeres, defined Z-lines, and intercalated disks, resembling the native heart tissue (Fig. 7.1D) [12].

The implementation of perfusion bioreactors in tissue engineering thus has greatly advanced the field, leading to the construction of functioning tissues with thickness much greater than seen before.

7.3 INDUCTIVE STIMULATION PATTERNS IN CARDIAC TISSUE ENGINEERING

7.3.1 MECHANOTRANSDUCTION AND PHYSICAL/MECHANICAL STIMULI

Mechanotransduction (biochemical response of cells to mechanical stimulation) is a very fundamental process in a living cell. The ability of cells to convert mechanical forces into biochemical regulatory information is crucial for normal physiology and pathological processes in many tissues. Mechanotransduction is mediated by various cellular and extracellular components, such as ECM components, cell-ECM adhesions (integrins, focal adhesions), cell-cell adhesions (cadherins and gap junctions), membrane components (ion channels, surface receptors), cytoskeletal components (microfilaments, microtubules), and, finally, intracellular and/or nuclear components (various kinases and transcription factors) that affect chromatin and gene expression [14, 15].

Cells of the myocardium are at home in one of the most mechanically dynamic environments in the body. Heart chamber filling and wall distention within diastole account for rapid changes in pressure and volume that are released by the wave of contraction that pumps blood through the body. At the cellular level, these pulsatile stimuli are experienced as cyclic strains (relative deformation) and stresses (force per unit area). Myocytes are stretched as the heart fills in diastole (preload) and actively contract against imposed load (afterload) in systole. Thus, mechanotransduction in cardiomyocytes occurs due to external tensile forces (outside in) and forces generated in cytoskeletons (inside out), including contractile forces that are produced through the sliding of actin and myosin filaments in sarcomeres [16, 17].

Cardiomyocytes employ numerous mechanisms to sense external forces and translate them into biochemical signals (Fig. 7.2). One of the most studied sensing/signaling pathways is ECM-integrin-cytoskeleton pathway. Binding of integrins on cell surface to ECM proteins promote assembly of "focal adhesion" anchoring complexes (though the activation of focal adhesion kinase and other signaling molecules). Focal adhesions play a central role in mechanotransduction. Force-dependent changes in shape and conformation of a subset of the load-bearing molecules in these adhesion plaques alter their biochemical activities, and thereby result in stress-dependent remodeling of the focal adhesion as well as associated changes in downstream signal transduction. Another mechanism of mechanical sensing is mediated by stretch-activated channels on the cell surface. These ion channels selectively regulate the permeability of the cell membrane to cations (Na^+, K^+, and Ca^{2+}) in response to strain, that, in turn, affects signaling and physiological response [16, 18, 19]. Different mechano-sensing mechanisms lead to activation of intracellular signaling pathways, most prominently cascades of G-proteins, mitogen-activated protein kinases (MAPK, including extracellular-regulated kinase (ERK), c-Jun N-terminal kinase (JNK), and the p38 MAP kinase), Janus-associated kinase/signal transducers and activators of transcription (JAK/STAT), protein kinase C (PKC), calcineurin, and intracellular calcium regulation (Fig. 7.2) [16, 19]. Multiple levels of cross-talk exist between these pathways.

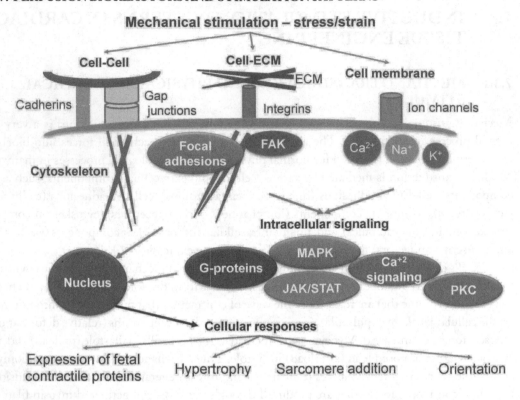

Figure 7.2: Major mechanotransduction signaling elements and responses. Stress and strain are sensed through various mechanisms, including cell-cell adhesions, cell-ECM interactions, and stretch-activated ion channels. This stimulation results in a series of intracellular events, including focal adhesion formation, cytoskeletal rearrangement and activation of various signaling cascades. These signals are translated to downstream responses, which culminates in changes in gene expression. The results of these changes include activation of fetal gene program, hypertrophy, sarcomere addition, and change in orientation and alignment. See text for more details. FAK, focal adhesion kinase; MAPK, mitogen-activated kinase; JAK/STAT, Janus-associated kinase/signal transducers, and activators of transcription; PKC, protein kinase C.

Cellular responses, that result from mechanical stimulation, are numerous and complex, and differ in kinetics and magnitude. Various parameters of stimulation, such as frequency, force magnitude and direction, are important factors that affect these responses, which are translated eventually to the physiological or pathological conditions. Mechanical stresses in the myocardium can be elevated by tissue damage following MI, excessive load (such as hypertension), or intrinsic defects, such as mutations in genes encoding contractile proteins. Cardiomyocytes respond to elevated loading conditions, in an attempt to adapt to new mechanical demands and restore wall stress to normal levels, with an increase in cell size, associated with transient and persistent molecular changes, a process generally referred as cellular hypertrophy. Changes in gene expression lead to an increased rate of contractile protein synthesis and cell mass. The long-term changes in gene expression also include the transition of some sarcomeric proteins to their fetal forms. Increase in protein mass lead to in-series or parallel addition of sarcomeres, depending on mechanical stress and overload nature (pressure or volume). Apart from changes in contractile protein mass, mechanical loading can also affect cytoskeletal or sarcomeric organization to regulate cell shape, alignment, and orientation. In pathological conditions, such as MI, hypertrophic response is initially beneficial, as it partially compensates for the decreased force generation and normalized wall stress. However, in the long term the hypertrophic response leads to cardiac failure. Local mechanics destabilization, altered calcium dynamics, structural disarray, and fibrosis are among the major factors that contribute to pathological remodeling and cardiac imbalance [17, 19].

The described processes and biomechanical dynamics are important in maturation and organization of native cardiac tissue. Thus, the ability of various engineering efforts to recreate those signals *in vitro* is critical for the formation of functional cardiac patches. As cardiac tissue is an excellent example of a tissue that constantly senses mechanostimulation *in vivo*, it may respond to a similar stimulation also *in vitro*.

The effect of mechanical stimulation on cells has been broadly investigated mainly in 2D cultivation systems. In 3D cell cultures, the homogenous distribution of these mechanical stimuli represents another challenge for bioreactor engineers. The most common examples for application of mechanical stimulation are bioreactors, developed to apply mechanical forces via piston/compression systems, substrate bending, hydrodynamic compression, and fluid shear [13, 20, 21, 22].

Zimmermann's group pioneered the application of mechanical stimulation to induce cardiac tissue engineering [2]. They applied a phasic mechanical stretch on previously casted circular molds, comprised of cardiac myocytes from neonatal rats, mixed with collagen I and matrix factors. In addition to the cyclic mechanical stretch, the medium in the culture system was constantly mixed to improve nutrient transfer, resulting in enhanced cell metabolic activity. The ring-shaped engineered heart tissue (EHT) developed by this approach displayed the important hallmarks of differentiated cardiac muscle tissue, i.e., striated myofibers and contractility.

Akhyari *et al* subjected gelatin-based cell constructs, seeded with human pediatric heart cells (non-contractile cardiomyocyte-like cells, isolated from ventricular tissue removed during surgical repair of Tetralogy of Fallot in children) to a cyclic stretch. They showed that this stimulation increased cell proliferation and distribution in the construct. In addition, collagen matrix formation and organization were enhanced by this stretch, as well as maximal tensile strength and resistance to stretch [1].

Mol and colleagues, when engineering a heart valve, *in vitro*, utilized a different stimulation system [23]. The stimulation pattern mimicked the diastolic phase of the cardiac cycle as well as the opening and closing behavior of the leaflet. A bioreactor system, developed for this purpose, applied a dynamic pressure difference over the tissue-engineered valve, and in doing so induced dynamic strains within the leaflets. Exposure of the heart valve construct to this strain regime has led to the development of a tissue-engineered aortic human heart valve for replacement.

Nowadays, a leading approach in mechanostimulation of the cardiac constructs is via stimulation during construct cultivation within a perfusion bioreactor. The advantages of such a bioreactor as a mechanostimulator are obvious; the interstitial medium flow within the porous cell construct also creates a frictional force on the surface of the cells, termed shear stress. Mammalian cells respond to fluid shear stress in different ways, for example by enhancing proliferation or changing morphology and organization.

Our group has used the perfusion bioreactor described in Section 7.2.3 to investigate the effect of pulsatile, interstitial fluid flow stimulation on cardiac tissue engineering. The various cellular events taking place in the cell construct, from the cell signaling levels up to the ultrastructural phenotype of the regenerated cardiac muscle tissue, were elucidated in this study [13]. When the bioreactor was operated at a shear stress of 0.6 dynes/cm^2, activation of the extracellular signal-regulated kinase (ERK) 1/2 signal transduction pathway was pronounced, followed by enhanced synthesis of the contractile proteins (Troponin T and sarcomeric α-actinin) and cell-cell interaction proteins (Cx-43 and N-cadherin). More importantly it was shown that long-term cultivation under the pulsatile flow regime like this promoted organization of thick, highly organized cardiac tissue, resembling the native heart. A model describing the proposed mechanism, translating mechanical cues to tissue regeneration, is presented in Fig. 7.3.

In a recent study, we described the micro-fabrication of a Multi-Shear Perfusion Bioreactor (MSPB), designed for cultivation of 3D cell constructs under differential fluid shear rates (Fig. 7.4) [24].

The MSPB device enables a simultaneous investigation of the effect of up to six different shear rates on the cell constructs. The multi-shear operation features of the bioreactor were modeled and then experimentally validated by measuring the effect of different shear rates on the behavior of human umbilical vein endothelial cells (HUVEC) seeded in macro-porous alginate scaffolds. Exposure of the constructs to three different levels of shear stress for 24 h resulted in constructs expressing three different levels of the membranal marker Intercellular Adhesion Molecule 1 (ICAM-1) and the phosphorylated endothelial nitric oxide synthetase (eNOS), markers known to be directly affected by shear stress. A longer period of cultivation, 17 days, under two different levels of shear stress, resulted in different lengths of cell sprouts within the constructs. Collectively, the HUVEC behavior within the different constructs confirms the feasibility of using the MSPB system for simultaneously imposing different shear stress levels, and for validating the flow regime in the bioreactor vessel as assessed by the computational fluid dynamic (CFD) model [24].

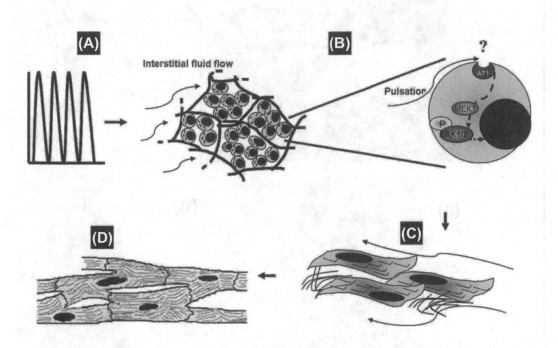

Figure 7.3: A proposed model describing the mechanistic effects of pulsatile interstitial fluid flow/shear stress on cardiac tissue regeneration. **A.** A pulsatile interstitial fluid flow, provided by the perfusion bioreactor subjects the scaffold-seeded cardiac cells to a physiological shear stress of 0.6 dynes/cm^2. **B.** The applied shear stress activates the mechanoreceptor, AT1, thus initiating the ERK1/2 cascade. In the presence of the MEK inhibitor, U0126, the pulsatile flow cannot activate ERK1/2. **C.** Activation of ERK1/2 cascade signals for cardiac cell survival and hypertrophy, as revealed by the enhanced synthesis of contractile cardiac proteins and proteins associated with cell-cell contacts and gap junction responsible for electrical coupling. **D.** With the secretion of collagen and its arrangement in fibers around the cardiac myofibers, the cardiac tissue reaches full maturity [13].

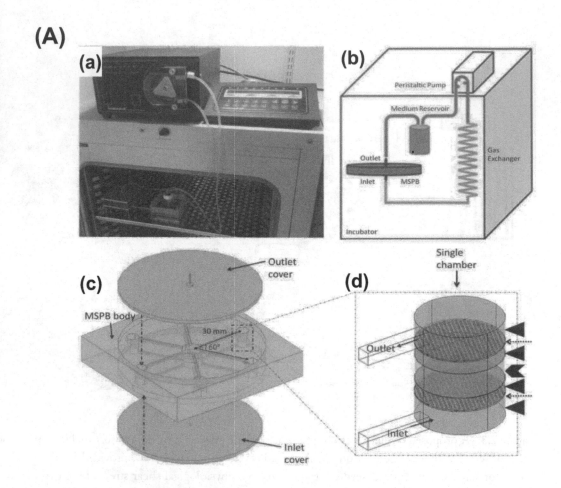

Figure 7.4: Multi-Shear Perfusion Bioreactor (MSPB) is designed for cultivation of 3D cell constructs under differential fluid shear rates. **A.** The MSPB design: (a) A picture of the MSPB system placed in the incubator; (b) The system and its flow circuit; (c) The bioreactor vessel has six chambers, located 30 mm from the center and connected via 22-mm long medium-leading channels to a medium flow splitter at the vessel center. The vessel is sealed by pressing top and bottom covers made of PDMS layer; (d) The chamber dimensions are 8.2-mm height and 6.1-mm diameter and thinner where the construct is held (5.9 mm). The cellular construct (red, indicated by a chevron) and the two flow-distributing nets (indicated by dashed arrows) placed 1 mm upstream and downstream to the construct, are held in place by porous plastic discs (blue, indicated by arrow heads). The direction of the medium flow is upward. Reprinted with permission from [24].

(B)

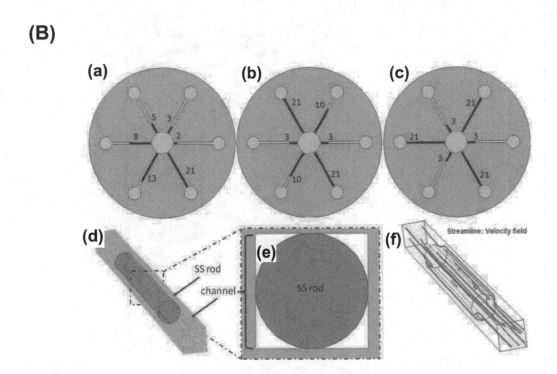

Figure 7.4: Multi-Shear Perfusion Bioreactor (MSPB) is designed for cultivation of 3D cell constructs under differential fluid shear rates. **B.** Illustration of the rods arrangement in the MSPB upper plane (as viewed from above) of three configurations of differential shear stress rates: The six-shear configuration (a) are utilized by inserting 2, 3, 5, 8, 13, and 21 mm long rods. Three shear configurations (b) are utilized by inserting 3, 10, and 21 mm long rods. Two shear configurations (c) are utilized by inserting 3 and 21 mm long rods. The exact same arrangement of rods was set on the other side of the bioreactor. Insertion of the SS rod to the channel blocks most of its cross-sectional area for a medium flow (d, e), forcing the medium to flow around it as demonstrated by the red streamlines (f). Reprinted with permission from [24].

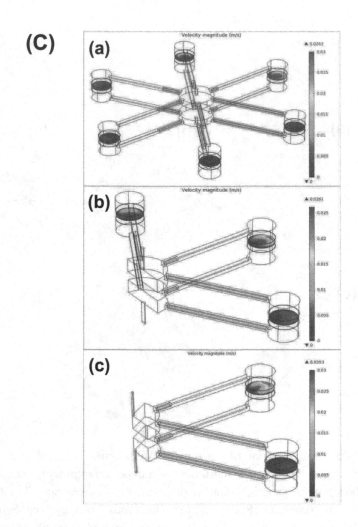

Figure 7.4: Multi-Shear Perfusion Bioreactor (MSPB) is designed for cultivation of 3D cell constructs under differential fluid shear rates. **C.** Computational model of fluid velocity in the bioreactor operated in: (a) six-shear manner by placing 6 different lengths of SS rods (2, 3, 5, 8, 13, and 21 mm) in the channels; (b) 3-shear manner by placing three different lengths of SS rods (3, 10, and 21 mm) in the channels; (c) 2-shear manner by placing two different lengths of SS rods (3 and 21 mm) in the channels. Reprinted with permission from [24].

7.3.2   MECHANICAL STIMULATION INDUCED BY MAGNETIC FIELD

The magnetic field has been attracting great interest as a tool for mechanical cell stimulation, although its signaling pattern within the stimulated cell is still not clear and poorly understood. Magnetic fields can protrude living cells since, unlike the electric field, they are not shielded by membrane potentials, and thus they can influence intracellular organelles. Also, due to their high penetration ability, magnetic fields can reach into deeper tissue layers when applied from a distance.

In stimulation settings, the magnetically mediated actuation "at distance" has a clear advantage compared to other stimulation setups since it enables highly controlled actuation both *in vitro* and *in vivo*. The target cell can be stimulated regardless of whether there are intervening structures, as long as these structures do not isolate the magnetic field. In addition, the magnetic field can be coupled with magnetically responsive particles that can be targeted to a specific cell or tissue site both *in vitro* and *in vivo*. Such coupling can easily and dynamically control the stress, applied directly by the magnetic particles to a desired area by varying the strength of the applied field.

Our group has recently developed magnetically responsive alginate scaffolds and tested their ability to provide means of a physical stimulation to living cells seeded within the scaffold (Fig. 7.5)

The nanocomposite alginate scaffolds were impregnated with magnetic iron-oxide nanoparticles (MNP), then seeded with bovine aortic endothelial cells, and the cell constructs has been exposed to an alternating magnetic field. The MNP-impregnated scaffolds were found to be more elastic compared to pristine scaffolds, while incorporation of MNP did not influence the macro-porosity structure of the scaffold (pore size and porosity) or their wetting extent by the culture medium. Endothelial cells cultivated within the magnetically-stimulated constructs showed significantly elevated metabolic activity during the stimulation period, most likely related to cell migration and re-organization into tube-like structures. Immunostaining and confocal microscopy examination on day 14 revealed that the magnetically-stimulated constructs, without supplementation of any angiogenic growth factors, contained vessel-like (loop) structures, while in the non-stimulated (control) scaffolds, the cells were mainly organized as sheets or aggregates (Fig. 7.5) [25].

It is still not clear what are the exact mechanisms acting on the cells within the nanocomposite scaffold under magnetic stimulation; however some speculations can be made. We showed that the impregnation process results in high density of MNP embedded within the scaffold wall, in close proximity and interacting with each other. Such direct NP interactions within the scaffold wall resemble the domain interactions existing in ferromagnetic material, leading to the phenomenon called magnetostriction. Magnetostriction causes the bulk materials to change their shape or dimensions during the process of magnetization, due to the domain interactions [26]. Such an effect would lead to overall scaffold contraction and consequently to cell stimulation, migration and organization into a tissue. This may explain the endothelial cell organization as observed in the magnetically-stimulated alginate scaffold. Although this hypothesis appears to be feasible, it needs further investigation and confirmation.

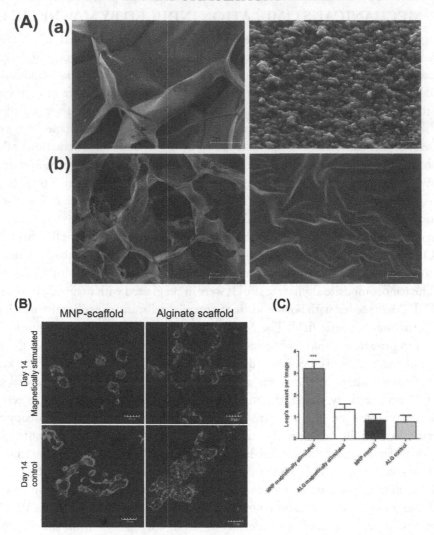

Figure 7.5: The promotion of *in vitro* vessel-like organization of endothelial cells in magnetically responsive alginate scaffolds. **A.** Scaffold morphology. Scanning electron microscopy (SEM) images of the (**a**) 1.4% (w/v) MNP-alginate and (**b**) non-magnetic alginate scaffolds. Note the presence of nanoparticles in MNP-alginate scaffold walls. **B.** Endothelial cell organization in MNP-alginate and alginate constructs, on day 14 post cell seeding. The cells are stained for F-actin (red) and nuclei (blue) (bar: 30 μm). **C.** Average loop number per image field counted on day 14 post cell seeding. A total of 25 randomly selected fields were analyzed per each group. Asterisks denote significant difference (by 2-way ANOVA), ***p < 0.005 (Bonferroni's pos-hoc test was used for comparison between the groups). Reprinted with permission from [25].

7.3.3 ELECTRICAL STIMULATION

From the moment it is formed, the heart acts as a sequential contracting syncytium. Although each cardiomyocyte is able to contract spontaneously, the overall heart function is controlled by a group of specialized pacemaker cells. These cells stimulate the generation of cardiac impulse and trigger synchronous contraction. In cardiac tissue engineering, the addition of exogenous electrical stimulation has attributed to the reconstruction of an appropriate environment for tissue regeneration by mimicking the pacemaker activity during heart tissue development.

Numerous studies explored the influence of electrical stimulation on cardiac and other cell types [27,28,29,30]. Studies performed on 2D cardiomyocyte cultures have revealed that the electrical stimulation (80–150 V, pulse duration of 5–10 ms, I < 5 mA, 1–5 Hz) induced cell enlargement, the development of more organized myofibrils, and greater expression of cardiac genes (ANF and MLC-2) [29]. Short-term electrical stimulation (60–120 min) of cardiomyocyte monolayers led to enhanced expression of Cx-43, an important gap junction protein, responsible for the mechanical and electrical communication between adjacent cells in the cardiac tissue [30].

Radisic and colleagues pioneered the application of electrical stimulation in cardiac cell constructs [3]. Electrical signals mimicking those in the native heart (rectangular pulses, 2ms, 5V/cm, 1Hz) were applied on neonatal cardiomyocytes seeded onto collagen constructs. Already after eight days of stimulation, the cardiac cells presented higher levels of gap junction proteins (Cx-43) expression compared to non-stimulated constructs. Electrical stimuli induced functional coupling, amplified contraction amplitude by a factor of 7, and there was also a significant level of ultrastructural differentiation [3]. Since these studies were performed under static conditions, the thickness of the engineered cardiac tissue was limited to 100 μm.

Additional works investigated the influence of electrical stimulation on endothelial cell behavior [31, 32]. Zhao *et al* reported that applied electric fields (EFs) of small physiological magnitude directly stimulate the production of vascular endothelial growth factor (VEGF) by endothelial cells in culture in the absence of any other cell type [32].

Recently, electrical stimulation has been implemented in tri-culture cell constructs (cardiomyocytes, endothelial cells, and fibroblasts) resulting in the formation of vascularized cardiac tissue [33].

Our group combined electrical stimulation with medium perfusion into one cultivation vessel to produce a thick functional cardiac patch (Fig. 7.6) [34].

For this, a custom-made electrical stimulator was integrated into the perfusion bioreactor described in Section 7.2.2. This was achieved by insertion of carbon rod electrodes between the two scaffold mesh holders. At first, the stimulation threshold for inducing a synchronous contraction in the cell constructs was determined under a microscope, by trial and error. Computer models of the electric fields (and current density) inside the bioreactor and the constructs were created. A successful stimulation of the cell construct in static cultivation mode was achieved at 6V with a current density of 74.4 mA/cm^2, while in the bioreactor, with the carbon rod electrodes, 5V was sufficient to achieve the same current density. Already after four days under a continuous electrical

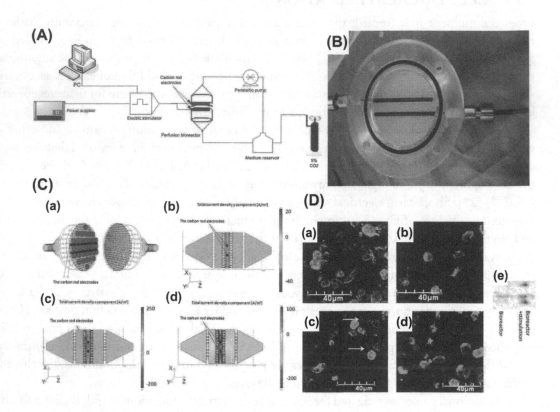

Figure 7.6: Electric field stimulation integrated into perfusion bioreactor for cardiac tissue engineering. **A.** Bioreactor setup. **B.** Configuration of the carbon electrodes integrated into the bioreactor. **C.** 3D electric field model in bioreactor. (**a**) An illustration of the 3D configuration with the carbon electrodes in black and the cell constructs in blue. (**b-d**) The current density in the y (**b**), z (**c**), and x (**d**) directions (1 μV between the electrodes). **D.** The effect of the electrical stimulation on cell morphology and Cx-43 levels. (**a, b**) Confocal microscopy images of anti-α-sarcomeric actinin immunofluorescence (green) of the cell constructs in the bioreactor with (**a**) or without (**b**) electrical stimulation. To-Pro (To-Pro 3 Iodide) was used for nuclear staining (red). The electrically-stimulated cell constructs are those placed in between the carbon electrodes. (**c, d**) Confocal microscopy images of anti-Cx-43 (red), anti-α-sarcomeric actinin (green) immunofluorescence of the cell constructs in a bioreactor with (**c**) or without (**d**) electrical stimulation. The white arrows indicate positive staining of Cx-43 between adjacent cells. To-Pro was used for nuclear staining (blue). The electrically-stimulated cell constructs are those placed in between the carbon electrodes. (**e**) Representative Western blot analysis for Cx-43 expression after four days of cultivation in the bioreactor with or without electrical stimulation [34].

stimulus in perfusion bioreactor, cell elongation and striation in the cell constructs were clearly detected, as well as enhanced expression level of Cx-43 [34].

7.4   SUMMARY AND CONCLUSIONS

This chapter described the development of bio-mimetic cell culture systems, designed to recapitulate some aspects of the actual *in vivo* cardiac environment. Advances in the design of perfusion bioreactors solved a critical problem in cardiac tissue engineering, leading to the enhancement of mass transfer within the construct and attaining thicker tissues. In conjunction with the applied mechanical and electrical stimulations, synchronously contracted cardiac patches were produced.

The field of *in vitro* cardiac tissue engineering has greatly advanced in recent years, as the cardiac patches developed so far mimic to a large extent the native cardiac tissue features and function. Clearly, the success of this strategy for scar replacement and /or support depends on the extent and how fast the integration into the host tissue would be. In the next chapter, we describe methods intended to increase vascularization of the cardiac patch, a process which should assist in its integration.

BIBLIOGRAPHY

[1] Akhyari P, Fedak PW, Weisel RD, Lee TY, Verma S, Mickle DA, et al. Mechanical stretch regimen enhances the formation of bioengineered autologous cardiac muscle grafts. Circulation. 2002;106:I137–42. DOI: 10.1161/01.cir.0000032893.55215.fc 87, 95

[2] Zimmermann WH, Schneiderbanger K, Schubert P, Didie M, Munzel F, Heubach JF, et al. Tissue engineering of a differentiated cardiac muscle construct. Circ Res. 2002;90:223–30. DOI: 10.1161/hh0202.103644 87, 95

[3] Radisic M, Park H, Shing H, Consi T, Schoen FJ, Langer R, et al. Functional assembly of engineered myocardium by electrical stimulation of cardiac myocytes cultured on scaffolds. Proceedings of the National Academy of Sciences of the United States of America. 2004;101:18129–34. DOI: 10.1073/pnas.0407817101 87, 103

[4] Radisic M, Yang L, Boublik J, Cohen RJ, Langer R, Freed LE, et al. Medium perfusion enables engineering of compact and contractile cardiac tissue. American journal of physiology. 2004;286:H507–16. DOI: 10.1152/ajpheart.00171.2003 87, 89

[5] Lanza RP, Langer RS, Vacanti J. Principles of tissue engineering. San Diego, CA: Academic Press; 2000. 88

[6] Radisic M, Park H, Gerecht S, Cannizzaro C, Langer R, Vunjak-Novakovic G. Biomimetic approach to cardiac tissue engineering. Philos Trans R Soc Lond B Biol Sci. 2007;362:1357–68. DOI: 10.1089/ten.2006.12.2077 89

[7] Radisic M, Marsano A, Maidhof R, Wang Y, Vunjak-Novakovic G. Cardiac tissue engineering using perfusion bioreactor systems. Nat Protoc. 2008;3:719–38. DOI: 10.1038/nprot.2008.40 89

[8] Shachar M, Cohen S. Cardiac tissue engineering, Ex-vivo: Design principles in biomaterials and bioreactors. Heart Fail Rev. 2003;8:271–6. DOI: 10.1023/A:1024729919743 89

[9] Freed LE, Guilak F, Guo XE, Gray ML, Tranquillo R, Holmes JW, et al. Advanced tools for tissue engineering: scaffolds, bioreactors, and signaling. Tissue Eng. 2006;12:3285–305. DOI: 10.1089/ten.2006.12.3285 89

[10] Carrier RL, Rupnick M, Langer R, Schoen FJ, Freed LE, Vunjak-Novakovic G. Perfusion improves tissue architecture of engineered cardiac muscle. Tissue Eng. 2002;8:175–88. DOI: 10.1089/107632702753724950 89

[11] Brown MA, Iyer RK, Radisic M. Pulsatile perfusion bioreactor for cardiac tissue engineering. Biotechnol Prog. 2008;24:907–20. DOI: 10.1002/btpr.11 89

[12] Dvir T, Benishti N, Shachar M, Cohen S. A novel perfusion bioreactor providing a homogenous milieu for tissue regeneration. Tissue Eng. 2006;12:2843–52. DOI: 10.1089/ten.2006.12.2843 89, 92

[13] Dvir T, Levy O, Shachar M, Granot Y, Cohen S. Activation of the ERK1/2 cascade via pulsatile interstitial fluid flow promotes cardiac tissue assembly. Tissue Eng. 2007;13:2185–93. DOI: 10.1089/ten.2006.0364 92, 95, 96, 97

[14] Ingber DE. Cellular mechanotransduction: putting all the pieces together again. Faseb J. 2006;20:811–27. DOI: 10.1096/fj.05-5424rev 93

[15] Sebastine IM, Williams DJ. The role of mechanical stimulation in engineering of extracellular matrix (ECM). Conf Proc IEEE Eng Med Biol Soc. 2006;1:3648–51. DOI: 10.1109/IEMBS.2006.260344 93

[16] Riehl BD, Park JH, Kwon IK, Lim JY. Mechanical Stretching for Tissue Engineering: Two-Dimensional and Three-Dimensional Constructs. Tissue Eng Part B Rev. 2012. DOI: 10.1089/ten.teb.2011.0465 93

[17] Curtis MW, Russell B. Micromechanical regulation in cardiac myocytes and fibroblasts: implications for tissue remodeling. Pflugers Archiv : European journal of physiology. 2011;462:105–17. DOI: 10.1007/s00424-011-0931-8 93, 95

[18] Parker KK, Ingber DE. Extracellular matrix, mechanotransduction and structural hierarchies in heart tissue engineering. Philos Trans R Soc Lond B Biol Sci. 2007;362:1267–79. DOI: 10.1098/rstb.2007.2114 93

[19] Lammerding J, Kamm RD, Lee RT. Mechanotransduction in cardiac myocytes. Annals of the New York Academy of Sciences. 2004;1015:53–70. DOI: 10.1196/annals.1302.005 93, 95

[20] Carver SE, Heath CA. Semi-continuous perfusion system for delivering intermittent physiological pressure to regenerating cartilage. Tissue Eng. 1999;5:1–11. DOI: 10.1089/ten.1999.5.1 95

[21] Guldberg RE. Consideration of mechanical factors. Annals of the New York Academy of Sciences. 2002;961:312–4. 95

[22] Roberts SR, Knight MM, Lee DA, Bader DL. Mechanical compression influences intracellular Ca2+ signaling in chondrocytes seeded in agarose constructs. J Appl Physiol. 2001;90:1385–91. 95

[23] Mol A, Driessen NJ, Rutten MC, Hoerstrup SP, Bouten CV, Baaijens FP. Tissue engineering of human heart valve leaflets: a novel bioreactor for a strain-based conditioning approach. Ann Biomed Eng. 2005;33:1778–88. DOI: 10.1007/s10439-005-8025-4 95

[24] Rotenberg MY, Ruvinov E, Armoza A, Cohen S. A Multi-Shear Perfusion Bioreactor for Investigating Shear Stress Effects in Endothelial Cell Constructs. Lab on a Chip. 2012. DOI: 10.1039/C2LC40144D 96, 98, 99, 100

[25] Sapir Y, Cohen S, Friedman G, Polyak B. The promotion of in vitro vessel-like organization of endothelial cells in magnetically responsive alginate scaffolds. Biomaterials. 2012. DOI: 10.1016/j.biomaterials.2012.02.037 101, 102

[26] Lee E. Magnetostriction and Magnetomechanical Effects. *Rep Prog Phys*. 1955;18:184. DOI: 10.1088/0034-4885/18/1/305 101

[27] Yamada M, Tanemura K, Okada S, Iwanami A, Nakamura M, Mizuno H, et al. Electrical stimulation modulates fate determination of differentiating embryonic stem cells. Stem Cells. 2007;25:562–70. DOI: 10.1634/stemcells.2006-0011 103

[28] Kawahara Y, Yamaoka K, Iwata M, Fujimura M, Kajiume T, Magaki T, et al. Novel electrical stimulation sets the cultured myoblast contractile function to 'on'. Pathobiology. 2006;73:288–94. DOI: 10.1159/000099123 103

[29] McDonough PM, Glembotski CC. Induction of atrial natriuretic factor and myosin light chain-2 gene expression in cultured ventricular myocytes by electrical stimulation of contraction. J Biol Chem. 1992;267:11665–8. 103

[30] Inoue N, Ohkusa T, Nao T, Lee JK, Matsumoto T, Hisamatsu Y, et al. Rapid electrical stimulation of contraction modulates gap junction protein in neonatal rat cultured cardiomyocytes: involvement of mitogen-activated protein kinases and effects of angiotensin II-receptor antagonist. J Am Coll Cardiol. 2004;44:914–22. DOI: 10.1016/j.jacc.2004.05.054 103

[31] Yue A, Yang G, Wu J, Lai Y, Huang H, Chen H. [The influence of the pulsed electrical stimulation on the morphology and the functions of the endothelial cells]. Sheng Wu Yi Xue Gong Cheng Xue Za Zhi. 2008;25:694–8. 103

[32] Zhao M, Bai H, Wang E, Forrester JV, McCaig CD. Electrical stimulation directly induces pre-angiogenic responses in vascular endothelial cells by signaling through VEGF receptors. J Cell Sci. 2004;117:397–405. DOI: 10.1242/jcs.00868 103

[33] Chiu LL, Iyer RK, King JP, Radisic M. Biphasic electrical field stimulation aids in tissue engineering of multicell-type cardiac organoids. Tissue Eng Part A.17:1465–77. DOI: 10.1089/ten.tea.2007.0244 103

[34] Barash Y, Dvir T, Tandeitnik P, Ruvinov E, Guterman H, Cohen S. Electric field stimulation integrated into perfusion bioreactor for cardiac tissue engineering. Tissue Eng Part C Methods. 2010;16:1417–26. 103, 104, 105

CHAPTER 8

Vascularization of Cardiac Patches

CHAPTER SUMMARY

Among the greatest challenges facing the implementation of cardiac patches as a successful therapeutic strategy for myocardial repair is how to vascularize the patch immediately after implantation and maintain its viability and function. The lack of proper vasculature in the engineered cardiac patch, as well as the slow and insufficient vascularization of the patch after implantation on the infarct zone, are the main causes for the failure of implanted *in vitro*-generated cardiac patches to improve infarct repair. In this chapter, we present an overview of the different strategies developed to induce vascularization of cardiac patches, including the use of the body as a bioreactor to induce rapid vascularization prior to implantation on the infarcted heart.

8.1 INTRODUCTION

Vascularization of the cardiac patch prior to its implantation on the infarcted myocardium is of a great challenge, and the realization of the cardiac patch strategy in clinical protocols depends on achieving this goal. Without immediate vascularization after implantation, the thick cardiac patches do not survive in the anoxic infarcted zone. Several strategies have been employed to induce vascularization of engineered cardiac patches. We categorize these strategies as direct and indirect ones. In direct approaches, the cell constructs include endothelial cells in addition to the functional cells. Indirect approaches include, for example, scaffold designs with discrete compartments for specific cell types and/or optimized for better vascularization upon implantation (see Section 6.6) [1, 2]. Other indirect approaches include cell seeding into the scaffold, with incorporation of proangiogenic growth factors (such as VEGF, PDGF-BB, and bFGF), either covalently immobilized to the scaffold or in the form of controlled delivery systems [3, 4, 5, 6, 7].

8.2 PREVASCULARIZATION OF THE PATCH BY INCORPORATING ENDOTHELIAL CELLS (ECS)

The most studied direct approach for *in vitro* patch pre-vascularization is the formation of a microvascular network within the engineered construct by co-culture of endothelial cells (ECs), embryonic fibroblasts (EmFs), and cardiomyocytes. Caspi *et al* reported the formation of synchronously contracting engineering human cardiac tissue in a co-culture of cardiomyocytes, ECs, and EmFs, derived from hESCs. The construct revealed the formation of endothelial vessel networks, stabilized by the presence of mural cells originating from the EmFs. Interestingly, the presence of the endothelial capillaries augmented cardiomyocyte proliferation and did not hamper cardiomyocyte orientation and alignment. Immunostaining, ultrastructural analysis (using transmission electron microscopy), RT-PCR, pharmacological, and confocal laser calcium imaging studies demonstrated the presence of cardiac-specific molecular, ultrastructural, and functional properties of the generated tissue constructs with synchronous activity mediated by action potential propagation through gap junctions [8]. Using a similar approach, Lesman *et al* developed an engineered human cardiac tissue by co-culture of hESC-derived CMs, EmFs, and human umbilical vein ECs (HUVECs). Transplantation of these constructs onto healthy rat hearts resulted in formation of both donor (human) and host (rat)-derived vasculature within the engrafted tri-culture tissue constructs, associated with functional integration of human-derived vessels with host coronary vasculature. The number of blood vessels was significantly greater in the tri-culture tissue constructs when compared with scaffolds containing only CMs. The performance of the constructs in infarcted myocardium remains to be examined [9].

Murry and co-workers used co-cultures of hESC- or iPSC-derived cardiomyocytes with HUVECs to engineer human cardiac muscle in 3D collagen matrices. Addition of HUVECs increased cardiomyocyte proliferation, and the further addition of stromal supporting cells (mouse embryonic fibroblasts and human marrow stromal cells) enhanced formation of vessel-like structures by ~10-fold. Subjection of the co-culture to a continuous mechanical load yielded a significantly more aligned tissue with good contractility. When transplanted onto healthy hearts of athymic rats, the engineered tissue formed stable grafts revealing the presence of human microvessels that were perfused by the host coronary circulation. The function of engineered tissue in myocardial repair after MI was not evaluated [10].

In all of the above studies, the three cell types- ECs, cardiomyocytes, and stromal cells, were seeded simultaneously into the scaffold and the culture medium used during cultivation was a mixture of the medium used for each individual cell. According to this, it may be possible that the seeded cells did not grow under optimal conditions.

Radisic *et al* have recently introduced a new strategy of seeding the cells sequentially into the scaffold. They showed that the sequential, but not simultaneous culture of ECs, fibroblasts (FB), and cardiomyocytes resulted in elongated, beating cardiac organoids [11]. In a follow-up paper, the group showed that the expression of Cx-43 and contractile function are mediated by VEGF released by the non-myocytes (EC, FB) during the preculture period (24 h) in the sequentially seeded cultures [12].

8.3   THE BODY AS A BIOREACTOR FOR PATCH VASCULARIZATION

A different approach to vascularize a tissue is by using the body as a continuous source of ECs and pericytes. The body is essentially used as a bioreactor, serving as a humidified, temperature-controlled incubator. This "bioreactor" can promote essential processes during tissue development, such as cell proliferation and differentiation. Lee and colleagues quantitatively compared the omentum, mesentery, and subcutaneous space as possible sites for the engraftment of an engineering tissue. According to their results, the highly vascularized omentum is the preferred tissue for patch engraftment and vascularization due to high secretion of growth and angiogenic factors, which promote angiogenesis and cell survival in the graft [13, 14, 15].

The body has been also utilized as an efficient vascularization tool in cardiac tissue engineering. Shimizu and colleagues noted that after transplantation of cardiac monolayers subcutaneously, vascularization within the constructs began promptly after implantation, creating a well-organized vascular network after a few days *in vivo* [16]. In another report, Suzuki *et al* showed that the omentum was able to maintain the viability and angiogenesis of transplanted myocardial cell sheets [17].

Our group has demonstrated the use of a peritoneal-generated cardiac patch to replace a full thickness of the ventricle in a rat model of heterotopic heart transplantation [18]. The patch was developed *in vitro* by seeding of fetal cardiomyocytes onto alginate scaffolds. Implantation of these patches into the rat peritoneum cavity resulted in the continuation of the cell and tissue development and extensive vascularization.

In a different study, we sought to use the rich blood vessel environment of the omentum to promote vascularization of a cardiac patch prior to its implantation on the infarct [19].

To enhance this process and better protect the cells within the patch, rat neonatal cardiomyocytes were seeded in scaffolds prepared from affinity-binding alginate, that were preloaded with a cocktail of pro-survival and angiogenic factors (insulin-like growth factor-1 (IGF-1), stromal cell-derived factor-1 (SDF-1), and VEGF). These scaffolds were previously shown to control multiple growth factor release and presentation (see Section 10.6). The incorporated SDF-1 and IGF-1 independently initiated activation of the Akt and ERK1/2 cardioprotective signaling pathways in the 3D environment. Collectively, the cocktail was able to maintain 100% cardiac cell viability during the short *in vitro* cultivation period prior to the patch transplantation onto the omentum. After the transplantation, SDF-1 within the patch was able to recruit BMSCs that differentiated to ECs and formed functional blood vessels (Fig. 8.1). Seven days post implantation on the omentum the patches were harvested, and after detecting the formation of proper networks of blood vessels the omentum-generated patches were re-transplanted onto the scar tissue of the infarcted rat hearts [19].

Four weeks after implantation, the omentum-generated patches were fully integrated into the host myocardium and showed thicker scars than those observed on the myocardium of control rats. The pre-vascularized patch was populated with striated and elongated cardiac cells that were positively stained for Troponin-T and Cx-43, indicating the formation of mechanical contacts between the transplanted cells (Fig. 8.1). The average relative scar thickness and blood vessel density in hearts treated with omentum-generated patches were statistically greater than the control groups, suggesting that these patches, with their full vascular content, thickened the scar. Importantly, omentum-generated patches were electrically coupled with the host myocardium, as assessed 4 weeks after engraftment using Langendorff-perfused isolated heart preparations. This integration

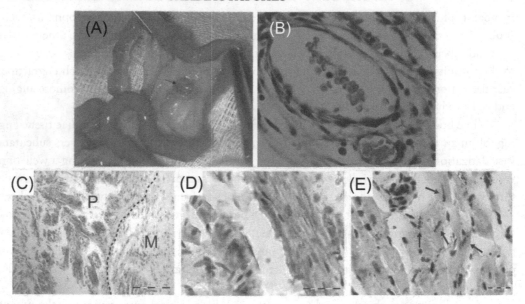

Figure 8.1: Prevascularization of a cardiac patch created in affinity-binding alginate scaffolds on the omentum improves its therapeutic outcome. **A.** The cardiac patch is transplanted on the omentum for seven days to promote its vasculogenesis. **B.** Functional blood vessels within the patch. **C.** H&E staining of cross-sections in the interface (dashed black line) of the host myocardium (M) and grafted omentum-generated patch (P), 28 days post transplantation. **D.** Typical cardiac cell striation could be observed by anti-Tn-T immunostaining (brown). **E.** Cx-43 expression (brown) between adjacent cardiomyocytes in omentum-generated cardiac patch suggests mechanical coupling. [Scale bar: 200 μm (C) and 20 μm (D and E)]

was evidenced by the higher amplitude of electrical signals in the scar zone and by the markedly lower capture threshold for pacing, indicating better excitability and/or electrical connectivity between the scar and healthy myocardium (Fig. 8.1, F-H). Finally, omentum-generated patches were able to preserve cardiac function and prevent LV dilatation and adverse remodeling, as shown by 2D echocardiography (Fig. 8.1, I-K).

Interestingly, similar beneficial results were obtained when the omentum-generated patch was constructed from an affinity-binding scaffold that was supplemented with only the mixture of pro-survival and angiogenic factors without seeded cardiac cells (Fig. 8.1) [19]. After regeneration on the omentum, these acellular constructs showed similar cell penetration and consistency as the cardiac-seeded constructs, which may explain their beneficial effects on infarct repair.

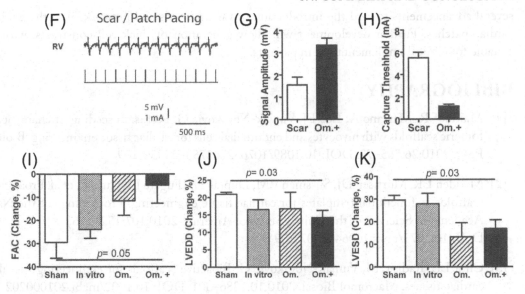

Figure 8.1: Prevascularization of a cardiac patch created in affinity-binding alginate scaffolds on the omentum improves its therapeutic outcome. **F.** Pacing through the scar electrode in hearts treated with an omentum-generated patch trigger synchronized beating of the healthy right ventricle, suggesting electrical integration of the patch. **(G)** Comparison of signal amplitude in the scar zone after grafting an omentum-generated patch (black) or in stitched hearts (white). **H.** Comparison of capture threshold intensity in hearts grafted with an omentum-generated patch (black) or only stitched (white). **I-K.** Cardiac function evaluation by 2D echocardiography of infarcted hearts after treatment with a stitch (sham), implantation of *in vitro*-grown patch, empty patch grown on the omentum (Om), or omentum-generated cardiac patch (Om$^+$): comparison of FAC **(I)** change, LVEDD **(J)**, and LVESD **(K)**. Changes were calculated as follows: [(values obtained after four weeks – baseline values)/baseline values] $\times 100\%$. Statistical evaluations were performed by unpaired t test and one-way ANOVA, P < 0.05. [19].

8.4   SUMMARY AND CONCLUSIONS

This chapter deals with a major issue in bioengineering of functional cardiac patches—the need for vascularization of the cell constructs before subsequent implantation into the infarct. One strategy developed to address this critical need relies on the addition of endothelial cells (ECs) to the cell mixture, to promote vessel formation in the developed tissue. Further refinement of this strategy includes sequential cell seeding and the use of different scaffold compartments (i.e., in channeled scaffolds) to improve cell organization and formation of primitive vascular networks. An alternative strategy described is the use of the body as a bioreactor for more effective pre-vascularization. Omentum-vascularized cardiac patches have shown improved electrical integration and beneficial effects on cardiac function in rat MI model. Collectively, the already available data clearly show several advancements toward the introduction of vascular beds and networks into the engineered cardiac patches. Further development will allow generation of thick cell constructs, with a size suitable for clinical implementation in patients.

BIBLIOGRAPHY

[1] Maidhof R, Marsano A, Lee EJ, Vunjak-Novakovic G. Perfusion seeding of channeled elastomeric scaffolds with myocytes and endothelial cells for cardiac tissue engineering. Biotechnol Prog. 2010;26:565–72. DOI: 10.1089/107632702753724950 109

[2] Madden LR, Mortisen DJ, Sussman EM, Dupras SK, Fugate JA, Cuy JL, et al. Proangiogenic scaffolds as functional templates for cardiac tissue engineering. Proceedings of the National Academy of Sciences of the United States of America. 2010;107:15211–6. DOI: 10.1073/pnas.1006442107 109

[3] Chiu LL, Radisic M, Vunjak-Novakovic G. Bioactive scaffolds for engineering vascularized cardiac tissues. Macromol Biosci. 2010;10:1286–301. DOI: 10.1002/mabi.201000202 109

[4] Miyagi Y, Chiu LL, Cimini M, Weisel RD, Radisic M, Li RK. Biodegradable collagen patch with covalently immobilized VEGF for myocardial repair. Biomaterials. 2011;32:1280–90. DOI: 10.1016/j.biomaterials.2010.10.007 109

[5] Saik JE, Gould DJ, Watkins EM, Dickinson ME, West JL. Covalently immobilized platelet-derived growth factor-BB promotes angiogenesis in biomimetic poly(ethylene glycol) hydrogels. Acta Biomater. 2011;7:133–43. DOI: 10.1016/j.actbio.2010.08.018 109

[6] Perets A, Baruch Y, Weisbuch F, Shoshany G, Neufeld G, Cohen S. Enhancing the vascularization of three-dimensional porous alginate scaffolds by incorporating controlled release basic fibroblast growth factor microspheres. J Biomed Mater Res A. 2003;65:489–97. DOI: 10.1002/jbm.a.10542 109

[7] Gao J, Liu J, Gao Y, Wang C, Zhao Y, Chen B, et al. A Myocardial Patch Made of Collagen Membranes Loaded with Collagen-Binding Human Vascular Endothelial Growth Factor Accelerates Healing of the Injured Rabbit Heart. Tissue Eng Part A. 2011. DOI: 10.1089/ten.tea.2011.0105 109

[8] Caspi O, Lesman A, Basevitch Y, Gepstein A, Arbel G, Habib IH, et al. Tissue engineering of vascularized cardiac muscle from human embryonic stem cells. Circ Res. 2007;100:263–72. DOI: 10.1161/01.RES.0000257776.05673.ff 110

[9] Lesman A, Habib M, Caspi O, Gepstein A, Arbel G, Levenberg S, et al. Transplantation of a Tissue-Engineered Human Vascularized Cardiac Muscle. Tissue Eng Part A. 2009. DOI: 10.1089/ten.tea.2009.0130 110

[10] Tulloch NL, Muskheli V, Razumova MV, Korte FS, Regnier M, Hauch KD, et al. Growth of engineered human myocardium with mechanical loading and vascular coculture. Circ Res. 2011;109:47–59. DOI: 10.1161/CIRCRESAHA.110.237206 110

[11] Iyer RK, Chiu LL, Radisic M. Microfabricated poly(ethylene glycol) templates enable rapid screening of triculture conditions for cardiac tissue engineering. J Biomed Mater Res A. 2009;89:616–31. DOI: 10.1002/jbm.a.32014 110

[12] Iyer RK, Odedra D, Chiu LL, Vunjak-Novakovic G, Radisic M. VEGF Secretion by Non-Myocytes Modulates Connexin-43 Levels in Cardiac Organoids. Tissue Eng Part A. 2012. DOI: 10.1089/ten.TEA.2011.0468 110

[13] Lee H, Cusick RA, Utsunomiya H, Ma PX, Langer R, Vacanti JP. Effect of implantation site on hepatocytes heterotopically transplanted on biodegradable polymer scaffolds. Tissue Eng. 2003;9:1227–32. DOI: 10.1089/10763270360728134 111

[14] Zhang QX, Magovern CJ, Mack CA, Budenbender KT, Ko W, Rosengart TK. Vascular endothelial growth factor is the major angiogenic factor in omentum: mechanism of the omentum-mediated angiogenesis. J Surg Res. 1997;67:147–54. DOI: 10.1006/jsre.1996.4983 111

[15] Zhou Q, Zhou JY, Zheng Z, Zhang H, Hu SS. A novel vascularized patch enhances cell survival and modifies ventricular remodeling in a rat myocardial infarction model. J Thorac Cardiovasc Surg. 2010;140:1388–96 e1-3. DOI: 10.1016/j.jtcvs.2010.02.036 111

[16] Shimizu T, Sekine H, Yang J, Isoi Y, Yamato M, Kikuchi A, et al. Polysurgery of cell sheet grafts overcomes diffusion limits to produce thick, vascularized myocardial tissues. Faseb J. 2006;20:708–10. DOI: 10.1096/fj.05-4715fje 111

[17] Suzuki R, Hattori F, Itabashi Y, Yoshioka M, Yuasa S, Manabe-Kawaguchi H, et al. Omentopexy enhances graft function in myocardial cell sheet transplantation. Biochemical and biophysical research communications. 2009;387:353–9. DOI: 10.1016/j.bbrc.2009.07.024 111

[18] Amir G, Miller L, Shachar M, Feinberg MS, Holbova R, Cohen S, et al. Evaluation of a peritoneal-generated cardiac patch in a rat model of heterotopic heart transplantation. Cell Transplant. 2009;18:275–82. DOI: 10.3727/096368909788534898 111

[19] Dvir T, Kedem A, Ruvinov E, Levy O, Freeman I, Landa N, et al. Prevascularization of cardiac patch on the omentum improves its therapeutic outcome. Proceedings of the National Academy of Sciences of the United States of America. 2009;106:14990–5. DOI: 10.1073/pnas.0812242106 111, 112, 113

CHAPTER 9

Acellular Biomaterials for Cardiac Repair

CHAPTER SUMMARY

An emerging body of evidence suggests that acellular forms of biomaterials have the capability to attenuate and prevent the deterioration in cardiac function after MI and instruct tissue repair. In this chapter, we describe the various acellular biomaterial strategies applied for inducing cardiac repair after MI in small and large animal models, from implantable porous scaffolds made from natural and synthetic polymers and decellularized organ-derived matrices, to injectable biomaterials, which present a more applicable clinical strategy. We then describe an alginate-based material, developed by our group, the first-in-man example of a biomaterial designed to be delivered as a solution via intracoronary injection and only at infarct to undergo gelation. This treatment has been proved to be safe in phase I/II clinical study. Finally, the last section speculates on the mechanism of biomaterial effects in cardiac repair.

9.1   INTRODUCTION

In the applications discussed so far, biomaterials have been used either as a vehicle for cardiac cell delivery into a damaged myocardium or as a platform for engineering of a functional cardiac patch. Although showing promising results in preclinical studies, the clinical implementation of the cellular-based strategies in patients still faces several important obstacles. A major one is the lack of a relevant cell source for transplantation. The various approaches for cell differentiation of pluripotent stem cells into cardiomyocytes still suffer from issues like low efficiency, ethical and/or technical difficulties, and safety concerns. The quest for the most suitable adult autologous stem cell source as a temporary alternative is also still ongoing.

Meanwhile, the potential of biomaterials as an acellular therapeutic strategy has been continuously expanded and established over the last decade or so. The biomaterials, initially developed for exogenous cell delivery and tissue engineering, are now investigated as a provisional supporting matrix for migrating host cells after implantation onto the damaged tissue.

In cardiac tissue engineering and regeneration, biomaterials have a great potential to play an especially critical role as compared to other organs. As mentioned in Chapter 2, adverse remodeling after MI has been associated with excessive damage to the cardiac ECM, which plays a critical role in the maintenance, integrity, and function of the heart tissue. These dynamic changes and alterations in cardiac ECM composition have a detrimental effect on both systolic and diastolic function [1, 2]. Biomaterials can be designed and developed to mimic cardiac ECM properties, such as chemistry, mechanical properties, morphology, hydrogel formation, and degradability. The achievements in the last decade or so provide a proof-of-concept for this strategy, by showing that biomaterials are capable of providing temporary tissue support, preserving cardiac function, facilitating self-repair, and inducing cardiac tissue regeneration.

9.2   ACELLULAR IMPLANTABLE SCAFFOLDS FOR *IN SITU* TISSUE SUPPORT

Implantable porous scaffolds have been investigated as a therapeutic strategy to thicken the infarct zone and by doing so to reduce the wall stress and attenuate LV remodeling after MI. When conceiving this strategy, it was hypothesized that the scaffold porous structure would enable its repopulation by infiltrating cells from the host, including vascular cells, and that after a complete integration of the patch to infarct zone, cardiac repair would be achieved. Since the inception of this strategy [3], several groups have investigated various types of porous scaffolds made of collagen or synthetic polymers [4].

Callegari *et al* implanted an acellular porous scaffold made of type I collagen on a rat infarcted heart, immediately after cryoinjury, and examined the degree of scaffold vascularization and tissue infiltration for up to 60 days [5]. The collagen scaffold was highly porous and with interconnected channels, and 60% of the material was resorbed within 60 days. The sponge attracted penetrating macrophages and induced strong angiogenic and arteriogenic responses at the infarct. In another study that also examined the effect of this strategy on functional activity of the heart, Goldman and co-workers used collagen type I scaffold (with a porosity of 70% and an average pore diameter of 30 to 60 μm) to repair non-transmural, not-ST-elevation MI induced by cryoinjury in rats. Three weeks after grafting, the scaffold that was integrated into the myocardium prevented adverse remodeling by reducing LV dilation, and increased angiogenesis. Six weeks after grafting, the scaffold was mostly absorbed by the underlying myocardial tissue [6].

Wagner and co-workers used an elastomeric biodegradable microporous polyester urethane urea (PEUU) scaffold for infarct repair in rats. This scaffold possessed 91% porosity and a 91 μm average pore size. The scaffold was placed and affixed on the infarct area two weeks after coronary artery ligation, and the outcome was analyzed after additional eight weeks. The patch region in the infarct was infiltrated with smooth muscle bundles with increased capillary density compared to control untreated animals. The treatment prevented LV dilation and improved cardiac function. These benefits were attributed to LV dilatation prevention, increased scar thickness, and reduced wall stress [7].

Although there are differences between the different studies, including the material in use and the extent of porosity and pore size, as well as different animal models, the strategy of implanting macroporous scaffolds to thicken the infarct zone has been proven to yield therapeutical benefits on cardiac repair.

9.3   DECELLULARIZED ECM

In parallel to the development of 3D matrices prepared from natural or synthetic materials, the emerging direction is the use of natural ECM as an appropriate matrix for cardiac tissue engineering through the process of organ decellularization. Decellularization is generally carried out by perfusion of the tissue with various detergents, aiming at cell removal without compromising the integrity and structure of the remaining ECM [8]. In such a way, the native structure and ECM composition is assumed to be preserved, and the matrix can be used for *in vitro* cell seeding or for instructive tissue restoration and support *in situ* [9, 10, 11]. The decellularized matrix can be lyophilized and milled to create a powder, which can be subsequently solubilized to create a liquid injectable matrix suitable for catheter delivery [12].

The main type of naturally derived decellularized ECM scaffold used for infarct or myocardial defects repair is prepared from porcine urinary bladder (UB). The ECM of this tissue represents a biologically latent membrane, which is already applied for various tissue engineering applications. Kochupura *et al* used this matrix for repair of full thickness defect in the right ventricle. Eight weeks after implantation, UB matrix improved regional systolic and diastolic function, compared to Dacron implantation. Importantly, cardiomyocytes were found in the ECM implant region, in contrast to Dacron implantation [9]. In another report, Badylak and co-workers evaluated the effect of UB matrix on infarcted hearts. At 6-8 weeks after the infarct, the matrix was placed as a full thickness LV wall replacement. Three months after implantation, myofibroblast infiltration associated with α-sarcomeric actin-positive cells was observed in the UB matrix group, compared to synthetic expanded polytetrafluoroethylene (ePTFE) patch. Cardiac function was not evaluated [13].

Ideally, the best source of a decellularized matrix for myocardial repair should be the matrix derived from the myocardium, which potentially possesses the composition and structure required for the specific needs of reconstructing the heart. The potential of such cardiac matrix for successful myocardial tissue engineering and bioartificial heart creation was proven in a proof-of-concept study performed by Ott *et al*, where whole rat hearts were decellularized and re-seeded with cardiac or endothelial cells. These hearts generated modest but detectable contractions and pump function [10]. Wainwright *et al* prepared intact cardiac ECM from whole porcine heart decellularization. This matrix supported the organization of chicken cardiomyocytes in a typical sarcomere structure, *in vitro* [14]. Eitan *et al* prepared cardiac ECM from LVs of porcine hearts, and showed the biocompatibility of the resulted scaffold for successful cell seeding, including cardiomyocytes, that showed functional phenotype and organization [11]. Mirsadraee *et al* prepared acellular matrix by decellularization of human pericardium. The matrix retained the major structural components and strength of the native tissue and showed biocompatibility when seeded with human dermal fibroblasts [15]. Although not matched to the properties of myocardial tissue, pericardium-derived matrices could represent an autologous option for tissue engineering applications aimed for myocardial support and repair.

Collectively, the use of decellularized matrices from natural ECM sources represents an attractive approach, as in this way the properties of the resulting scaffold may potentially be matched to the target tissue, including the heart. However, several technical issues are still to be resolved, such as the need for effective decellularization protocols, possible immunogenicity, preservation/storage, etc.

9.4 INJECTABLE BIOMATERIALS

9.4.1 INJECTABLE HYDROGELS BASED ON NATURAL OR SYNTHETIC POLYMERS

Various natural or synthetic polymers can form hydrogels, namely, solid networks of physically or chemically cross-linked polymer chains with varying water content, which can be directly injected into the infarcted heart. Such injectable hydrogels can recapitulate the microenvironment for inducing effective tissue repair, by replacing damaged ECM, providing temporary support for the infarct, and instructing tissue restoration. These effects could prevent the adverse remodeling and progression of heart failure. The application of injectable biopolymers is less invasive than implanting macroporous scaffolds, and is, therefore, more clinically appealing. Indeed, various natural or synthetic acellular injectable biomaterials have been found to improve the outcome after MI [16, 17, 18].

Christman *et al* pioneered the field of injectable biomaterials by exploring the effects of fibrin glue as a treatment strategy for MI. Fibrin forms a crosslinked 3D hydrogel in the myocardium upon injection with a dual-barreled syringe. One barrel contains fibrinogen and aprotinin (a fibrinolysis inhibitor), and the second barrel contains thrombin, factor XIIIa, and $CaCl_2$. Following a similar mechanism to that involved in the normal clotting cascade *in vivo*, when fibrinogen and thrombin are mixed, fibrinogen is converted to fibrin which self-assembles and is crosslinked via the factor XIIIa. Fibrin hydrogel injection, in the ischemia/reperfusion (I/R) model, preserved cardiac function and scar thickness, five weeks after injection [19]. In another study, fibrin injection was associated with reduced infarct size, increased angiogenesis (shown by the increase in microvessel density), compared to BSA injection [20]

Utilizing a large animal swine model, Mukherjee *et al* injected composite hydrogels containing both fibrin and alginate to prevent geometric LV remodeling. One week post-MI, 200 μL each of the 25 injections were applied to the infarct area via a double-barreled injection device; one component was comprised of fibrinogen, fibronectin, factor XIII, plasminogen, and gelatin-grafted alginate dissolved in an aprotinin solution, while the second consisted of the cross-linking agents, thrombin and $CaCl_2$. The therapeutic outcomes of this treatment included increased posterior wall thickness 1 week postinjection and a reduction in infarct expansion 21 and 28 days post-MI; however, no functional improvements were observed [21]

Yu *et al* investigated and compared the therapeutic effects of fibrin and alginate in a rodent model of chronic ischemic cardiomyopathy. The researchers found that both polymers can augment left ventricular wall thickness, resulting in reconstruction of left ventricular geometry and improvement of cardiac function. Echocardiography results at five weeks after injection of alginate demonstrated persistent improvement of left ventricular fractional shortening (FS) and prevention of a continued enlargement of left ventricular dimensions, whereas fibrin glue demonstrated no progression of left ventricular negative remodeling. There was increased arteriogenesis in both the alginate and fibrin glue groups compared with that seen in the phosphate-buffered saline control group. Infarct size was significantly reduced in the fibrin group, and there was a trend toward smaller infarcts in the alginate group.

Chitosan is a linear polysaccharide that is biocompatible and biodegradable and therefore has been used in a wide variety of tissue engineering applications. Chitosan hydrogels can be formed upon mixing of commercially produced chitosan with a glycerol phosphate and glyoxal solution. These gels exhibit a thermoresponsive gelation that is tuned to occur at 37°C by changing the glyoxal concentration, while hydrogel degradation is controlled by the degree of deacetylation [18]. In a rat infarct model, a thermally responsive chitosan was injected one week post-MI [22]. Four weeks after hydrogel injection, the myocardium thickness was significantly increased compared to PBS controls, even though the amount of chitosan presented in the myocardium after four weeks had substantially decreased due to hydrogel degradation. There were also significant improvements in infarct size, FS, EF, end systolic diameter (ESD), end diastolic diameter (EDD), and microvessel density.

Collagen is a natural ECM protein that has been applied for LV remodeling therapies due to the ability to inject as a liquid, which subsequently gels at 37°C [18]. Collagen injections (95% type I, 5% type III) in one week-old rat infarcts substantially increased infarct thickness, stroke volume (SV), and EF compared to saline injection controls; there was also a trend for smaller end systolic volume (ESV) and larger end-diastolic volumes (EDV) in biomaterial-treated animals [23].

Hyaluronan (HA) is a natural un-sulfated polysaccharide that is abundant in the body and plays a major role in several biological processes that include angiogenesis cell migration, and it is also a lubricant material. In most, if not all of the tissue engineering related strategies, only modified HA has been used; the modifications used to improve the material mechanical strength and adding functional groups for inducing photopolymerization leading to scaffold formation [18]. A study by Ifkovits *et al* tested therapeutics effects of an injectable hydrogel made from modified HA (methacrylated HA (MeHA)) with a high (43 kPa) and low (7.7 kPa) compression moduli in an ovine model of MI [24]. Treatment with both hydrogels significantly increased the wall thickness in the apex and basilar infarct regions compared with the control infarct. However, only the higher-modulus (MeHA High) treatment group had a statistically smaller infarct area compared with the control infarct group. Moreover, reductions in normalized end-diastolic and end-systolic volumes were observed for the MeHA High group. This group also tended to have better functional outcomes (cardiac output and EF) than the low-modulus (MeHA Low) and control infarct groups. This study emphasizes the importance of the rational material design and mechanical properties on therapeutic outcome post-MI.

Synthetic materials provide additional potential in engineering a variety of gelation mechanisms and physical properties. One synthetic thermosensitive polymer, comprised of dextran (Dex) grafted poly(caprolactone)-2-hydroxyethyl methacrylate (PCL-HEMA) and copolymerized with poly(N-isopropylacrylamide) (PNIPAAm) termed Dex-PCL-HEMA/PNIPAAm, was developed to gel *in situ*. MI was induced in rabbits by coronary artery ligation; four days later, 200 μL Dex-PCL-HEMA/PNIPAAm of gel solution was injected into the infarcted myocardium. Injection of phosphate-buffered saline served as control. Thirty days after treatment, histological analysis indicated that injection of the biomaterial prevented scar expansion and wall thinning compared with controls. Echocardiography studies showed that injection of hydrogel increased left ventricular ejection fraction and attenuated left ventricular systolic and diastolic dilatation. Haemodynamic analysis demonstrated improved cardiac function following implantation of the hydrogel [25]

9.4.2   INJECTABLE DECELLULARIZED ECM MATRICES

Decellularized ventricular and pericardial ECM can be processed to create a solubilized liquid with the ability to gel via self-assembly at physiological temperature both *in vitro* and *in vivo* upon injection into myocardial tissue [26, 27]. For this purpose, ventricular or pericardial tissue was harvested and decellularized using sodium dodecyl sulfate detergent. The decellularized matrix was then lyophilized and milled to create a fine powder, which was then solubilized using enzymatic digestion to create a liquid matrix for catheter-based delivery. The decellularized ventricular ECM biochemical composition was shown to retain the complexity of proteins, peptides, and glycosaminoglycans of natural ECM. Initial *in vivo* feasibility testing showed that the solubilized myocardial matrix undergoes gelation in healthy rat myocardial tissue upon direct epicardial injection. The *in situ*-formed hydrogel had a pore size of $\sim$30 μm and its nano-fibrillar structure resembled that of the intact decellularized tissue, prior to processing. The myocardial decellularized matrix was shown to promote the migration of human coronary artery endothelial cells and rat aortic smooth muscle cells *in vitro* as well as to promote the infiltration of vascular cells and the formation of arterioles *in vivo* [27]. The researchers also created pig and human pericardium-derived decellularized ECM injectable matrices by a similar procedure. The matrices induced neovascularization and promoted cell mobilization, suggesting that this type of constructs can represent an injectable scaffold for cardiac tissue engineering. Moreover, as the pericardium can be surgically resected by minimally invasive thoracoscopic pericardiectomy, this tissue layer represent a potentially autologous source of scaffold material, while sacrificing the properties of ECM derived from myocardium [26].

In a recent study, using a similar fabrication procedure, Syngelin *et al* prepared an injectable myocardial hydrogel from decellularized porcine ventricular ECM [28]. Rats underwent ischemia/reperfusion followed by intramyocardial injection of the hydrogel or saline two weeks later. Immunohistochemical assessment showed an increase in the size of cardiomyocyte islands surviving in the infarct area of the hydrogel-treated animals, suggesting that the matrix injection may act to salvage remaining cardiomyocytes in the infarct. The injection of myocardial matrix hydrogel was also associated with preserved cardiac function, as shown by preserved values of EF, end-diastolic and systolic volumes (evaluated by MRI). The authors also assessed arrhythmia inducibility using programmed electrical stimulation *in vivo* by burst and extra stimulus pacing protocols in rats one week post-injection of myocardial matrix compared to saline. *In vivo* pacing protocols did not show statistical difference when comparing the average incidence of ventricular tachycardia between hydrogel and saline-treated animals. Finally, the clinical feasibility of the myocardial hydrogel injection in healthy and infarcted pig hearts was shown by using minimally invasive percutaneous transendocardial catheter, under the guidance of unipolar electromechanical mapping (NOGA). A biotinylated matrix was delivered into 15 injection sites in the infarct and border zones, with no signs of pericardial effusion. Biotinylated matrix staining confirmed the presence of the material in the myocardial wall, as well as gelation of the matrix *in vivo*; the presence of the matrix was not observed in the satellite organs, confirming the lack of leakage of the material into the ventricle [28]. Collectively, this study demonstrates the feasibility and efficacy of decellularized ventricular ECM-derived *in situ*-forming hydrogel for the treatment of MI. These promising results await confirmation in a large animal model.

9.5   FIRST-IN-MAN TRIAL OF INTRACORONARY DELIVERY OF ALGINATE BIOMATERIAL

Our group developed an injectable alginate biomaterial which can be delivered by intracoronary injection as a solution. At the infarct, due to the high calcium ion concentration immediately after acute MI, the solution undergoes gelation forming a hydrogel [29, 30].

The injectable solution is a partially cross-linked alginate network, prepared by mixing a 1% (w/v) solution of 30-50 kDa sodium alginate (having high G to M ratio), and 0.3% (w/v) D-gluconic acid/hemicalcium salt, and is capable of flowing due to its relatively low apparent viscosity ($\sim$10cP) [31]. The unique mechanical properties of the partially cross-linked alginate solution were exemplified by rheology, the mechanical spectra revealing that the storage (G') and loss (G") moduli of the solution are closely related or sharing a cross-point. This type of physical behavior usually characterizes cross-linked material in the verge of phase transition from its liquid state into a hydrogel, and such transition can occur by increasing local cation concentration [31].

At the acute infarct, the partially cross-linked alginate solution undergoes a rapid gelation and phase transition into a hydrogel due to the additional cross-linking, in response to the elevated concentrations of calcium ions at the infarct after MI and due to water diffusion from the injectable solution to the surrounding tissue (Fig. 9.1A) [29, 30, 31]. The alginate hydrogel degrades and disappears from the infarct zone with time; 6 weeks after administration into the infarct, only remnants of biotinylated alginate material remained at infarct (Fig. 9.1B) and it was replaced by a host tissue composed of myofibroblasts and enriched with blood capillaries (Fig. 9.1C). The alginate hydrogel dissolution occurs via an exchange reaction between the crosslinking calcium ions by sodium ions from the surrounding tissue; a process occurring with time at the healing infarct due to the reduction in calcium ion concentration (Fig. 9.1A):

$$2NaAlg + Ca^{2+} \rightleftarrows 2Na^+ + Ca(Alg)_2$$

The beneficial therapeutic effects of this novel *in situ*-forming alginate hydrogel on MI repair have been proven in recent and chronic models of MI in rats and in acute model in pigs [29, 30]. In an acute MI model in rats, myocardial injection of the partially cross-linked alginate solution into

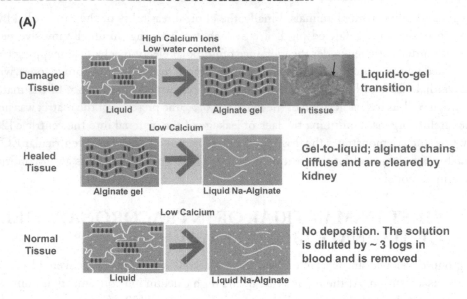

Figure 9.1: Beneficial therapeutic effects of injectable alginate biomaterial on LV remodeling after MI in rats (**B-E**) and pigs (**G-I**). **A.** A schematic model describing the three possible occurrences after intracoronary injection of the partially cross-linked alginate solution. In a damaged tissue after acute MI, the partially cross-linked alginate solution undergoes gelation at the infarct due to elevated calcium ion concentration. During healing, calcium ion concentration decreases in the tissue, leading to hydrogel dissolution via an exchange reaction of calcium ions with sodium ions. The water-soluble alginate molecules can diffuse and be excreted via the urine due to their low molecular weight (less than 50 kDa). In normal healthy tissue, no gelation occurs, and the solution is diluted in systemic circulation and is removed via urine. Reprinted with permission from [29, 30].

the infarct zone resulted in increased scar thickness and improved LV dimensions compared with saline, two months after injection. These results were similar or superior to injection of suspension of neonatal cardiomyocytes (Fig. 9.1C-D).

In a large animal model, we showed that intracoronary injection of the partially cross-linked alginate solution into the infarcted hearts of pigs is feasible and safe. The deposition of alginate biomaterial at the infarct zone (Fig. 9.1E) prevented and even reversed LV enlargement and increased scar thickness by 53% compared with saline, two months after the injection (Fig. 9.1F) [30]. These beneficial effects of the alginate hydrogel were dose-dependent (Fig. 9.1G) and are likely due to temporary replacement of the functions of the damaged ECM, followed by increased cellular infiltration during hydrogel erosion (Fig. 9.1H) [29, 30]. The *in situ*-forming alginate hydrogel was shown to undergo slow erosion and dissolution over a period of six weeks, during healing of the infarct.

(B)

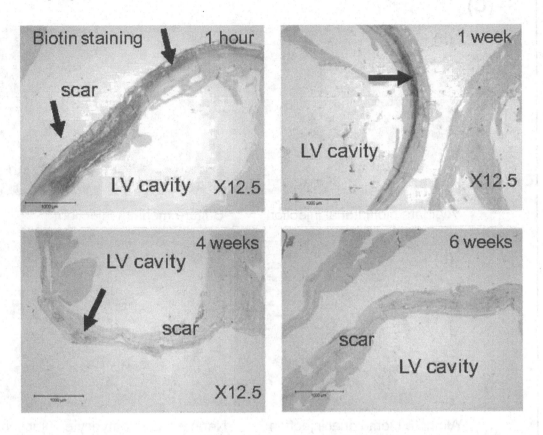

Figure 9.1: Beneficial therapeutic effects of injectable alginate biomaterial on LV remodeling after MI in rats (**B-E**) and pigs (**G-I**). **B.** The deposited alginate hydrogel dissolves and disappears from the infarct within six weeks. Distribution of biotinylated-labelled alginate biomaterial at the infarct after intramyocardial injection of calcium cross-linked alginate solution, in rat. The areas of positive biotin staining, as a percentage of the scar area, were decreased significantly at 1, 4, and 6 weeks after injection. Reprinted with permission from [29, 30].

These encouraging results have led to a first-in-man clinical trial, proving the safety of intracoronary injection of alginate biomaterial in acute MI patients [32]. Due to the localized effect of the biomaterial, it received FDA approval for a regulatory path as a medical device, intended to be injected to patients following acute myocardial infarction, for prevention of ventricular remodeling

(C)

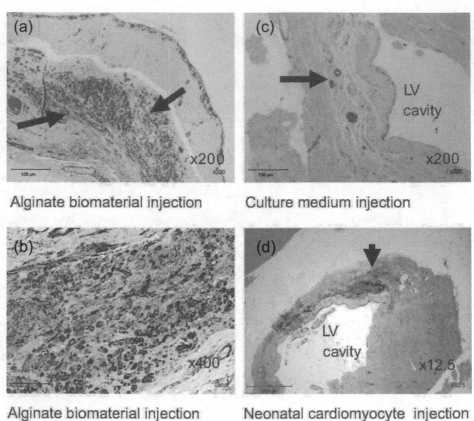

Figure 9.1: Beneficial therapeutic effects of injectable alginate biomaterial on LV remodeling after MI in rats (**B-E**) and pigs (**G-I**). **C**. Representative micrographs of infarcted hearts treated with biomaterial or cardiomyocyte transplantation (immunostaining for α-smooth muscle actin (SMA)). (**a**) Examination of the scar tissue 8 weeks after alginate biomaterial injection revealed extensive positive brown staining (arrows). (**b**) Higher magnification showed that the biomaterial-treated scar is populated with numerous SMA-positive cells, probably myofibroblasts, and with small vessels (marked by arrows). The scaffold-treated scars were thicker than the scars treated with culture medium (**c**). **d**, Neonatal cardiac cell implant (arrow) at the border of the infarct zone, 8 weeks after cell suspension injection. Engrafted cells appeared yellow-brown, were undifferentiated, and were isolated from the host myocardium. Reprinted with permission from [29, 30].

(D)

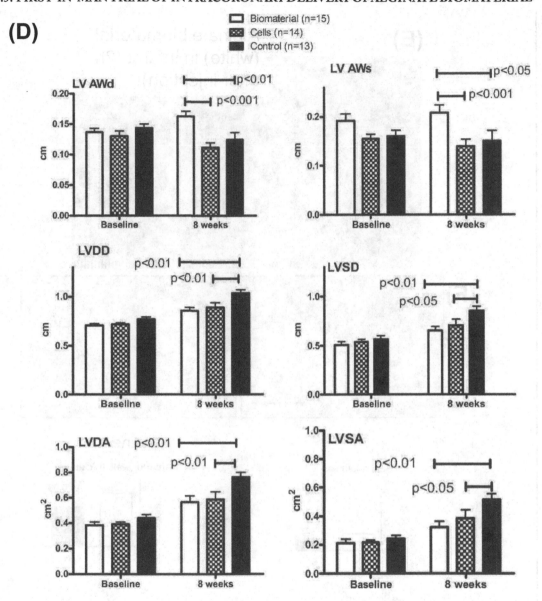

Figure 9.1: Beneficial therapeutic effects of injectable alginate biomaterial on LV remodeling after MI in rats (**B-E**) and pigs (**G-I**). **D.** Comparison of the functional effects of alginate biomaterial vs. cardiomyocytes vs. saline injection into recent (seven-day-old) scars in rats. Individual values and mean (±SE) are shown. The probability values are Bonferroni-adjusted for three comparisons. AWd indicates anterior wall diastolic thickness; AWs, anterior wall systolic thickness; LVDD, LV end-diastolic dimension; LVSD, LV end-systolic dimension; LVDA, LV end-diastolic area; and LVSA, LV end-systolic area. Reprinted with permission from [29, 30].

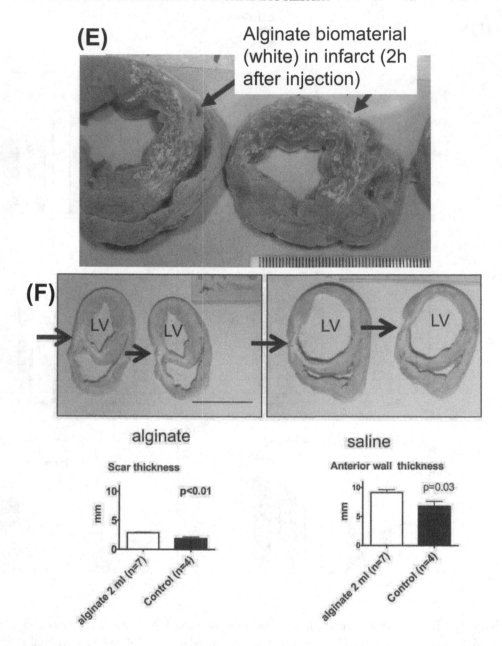

Figure 9.1: Beneficial therapeutic effects of injectable alginate biomaterial on LV remodeling after MI in rats (**B-E**) and pigs (**G-I**). **E and F.** Reprinted with permission from [29, 30]. *(Continues. Caption on the next page.)*

Figure 9.1: *(Continued.)* **E.** Macroscopic views of heart sections, 2 hours after intracoronary injection of calcium cross-linked alginate solution, reveals the *in situ* deposition of the alginate hydrogel (white areas, arrows), nicely distributed in the infarct. **F.** Post-mortem morphometry at 60 days after MI indicates that the alginate implant increases scar and arterial wall thicknesses as compared to saline treatment. Representative sections of heart treated with alginate (upper left panel) or saline (upper right panel). Morphometry shows that injection of alginate solution (2 ml) increases scar thickness (arrows) as well as anterior wall thickness (lower panels).

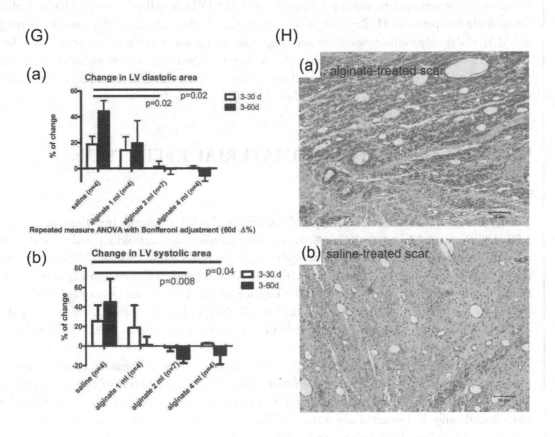

Figure 9.1: Reprinted with permission from [29, 30]. *(Continues. Caption on the next page.)*

and subsequent congestive heart failure. At present, the PRESERVATION I pivotal clinical trial is running for the material, now known as Bioabsorbable Cardiac Matrix (BCM).

Figure 9.1: *(Continued.)* Beneficial therapeutic effects of injectable alginate biomaterial on LV remodeling after MI in rats (**B-E**) and pigs (**G-I**). **G.** The effect of intracoronary injection of various volumes of alginate solution on LV dilation, 30 and 60 days after MI in pigs. Comparison of the therapeutic effects of intracoronary delivery of 1, 2, and 4 ml of alginate solution or saline (2 ml) into recent (4-day-old) scar. All volumes of alginate attenuate or prevent left ventricular (LV) diastolic (**a**) and systolic (**b**) dilation compared with control. Relative change was calculated as ([30 or or 60 day measure—3 day measure]/3 day measure) ×100. Individual values and mean (±SEM). The p values are for treatment effect versus control by repeated-measures analysis of variance (ANOVA) and Bonferroni post hoc test adjusted for multiple comparisons. **H.** Representative micrographs of infarcted hearts after immunostaining for α-SMA, 60 days after either alginate or saline injection. Examination of the scar tissues shows that (**a**) the alginate-treated scar is populated with numerous blood capillaries and myofibroblasts that are stained positive for α-SMA and (**b**) the saline treated scars show positive staining for α-SMA predominantly on the vessel walls. Reprinted with permission from [29, 30].

9.6 MECHANISM OF BIOMATERIAL EFFECTS ON CARDIAC REPAIR

The possible mechanisms behind the beneficial effects of sole biopolymer injection are most likely related to the increase in scar thickness, early infarct stabilization, scaffolding, and critical physical support to the healing of LV, as well as replacement for the damaged ECM. All these effects are significant for reducing wall stress, prevention of LV dilatation, effective healing and repair. By thickening the scar, wall stress is reduced (by Laplace law) and the degree of outward motion of the infarct that occurs during systole (paradoxical systolic bulging) is also reduced. This is a significant effect, since one of the most important predictors of mortality in patients with MI is the degree of LV systolic dilatation.

The functional improvement seen after biomaterial treatment of the infarct was not accompanied by an actual induction of tissue regeneration, meaning without addition of new contractile units. This passive type of mechanical regeneration was confirmed by utilizing computational simulation models analyzing the impact of any material (ECM-like materials and/or cell masses) injection into infarcted myocardium [33]. Using a finite element (FE) model to simulate the effects of injecting a non-contractile material into the myocardium, Wall *et al* showed that bulking the myocardium was sufficient to attenuate post-MI geometric changes and, thus, to decrease stress in the myocardial wall. More specifically, they demonstrated that injections of 4.5% of the LV wall volume and 20% of the stiffness of the natural myocardium into the infarct border zone were able to decrease the fiber stress by 20% compared to control simulations with no injections [33].

9.7 IMMUNOMODULATION OF THE MACROPHAGES BY LIPOSOMES FOR INFARCT REPAIR

Biomaterials mimicking natural processes after MI may be used to influence infarct repair. In this section, we present a new biomaterial-based strategy for the modulation of cardiac macrophages to a reparative state, at a predetermined time after MI, in aim to promote resolution of inflammation and elicit infarct repair.

9.7.1 INFLAMMATION, APOPTOSIS, AND MACROPHAGE RESPONSE AFTER MI

Inflammation has emerged as a critical biological process contributing to nearly all aspects of cardiovascular diseases including MI. After MI, cardiomyocyte death triggers a cascade of inflammatory pathways. For the heart, an organ with limited regeneration ability, post-MI remodelling comprises a series of structural and functional changes, all of which have been linked to the activation of the inflammatory pathways. Inadequate or excessive inflammatory response may lead to improper cellular repair, tissue damage, and dysfunction. Thus, the modulation of inflammation as a therapeutical strategy for cardiac repair after MI has become an active research field.

Our group focussed on macrophage modulation as a strategy for inducing cardiac tissue repair. After MI, resident and recruited macrophages remove necrotic and apoptotic cells, secrete cytokines, and modulate angiogenesis at the infarct site. As such, the macrophage is a primary responder cell involved in the regulation of post-MI infarct wound healing at multiple levels [34]. Apparently, different macrophage populations are responsible for these activities; during the early inflammatory phase (phase 1), proinflammatory macrophages dominate the injury site and phagocytose the apoptotic/necrotic myocytes, whereas during inflammation resolution (phase 2), the dominant macrophage population is the reparative macrophages, which promote infarct repair [35, 36]. The duration and extent of the early inflammatory phase have major implications on infarct size and ventricular remodelling.

Thus, we hypothesized that immunomodulation to control the duration of the different inflammation phases following MI could be a therapeutic target to improve infarct healing and repair. Until our study, most, if not all, investigated immunomodulation strategies for myocardial repair relied on the delivery of various growth factors or activated cells (macrophages, apoptotic cells, stem cells) [37, 38]. Such approaches are less clinically relevant, less safe, and certainly not as reproducible and accessible as an acellular system, and therefore have less potential to serve as a treatment for inflammation associated with MI in patients.

9.7.2 PS-PRESENTING LIPOSOMES AS APOPTOTIC CELL-MIMICKING PARTICLES FOR EFFECTIVE IMMUNOMODULATION AND INFARCT REPAIR

We conceived a novel strategy for controlling the duration and extent of the inflammatory phase following MI, in aim to reduce myocardium damage, preserve infarct size, and prevent ventricle remodeling. Our strategy exploits the principle underlying the anti-inflammatory effects of apoptotic cells, which are known to actively suppress inflammation by inhibiting the release of proinflammatory cytokines from macrophages while augmenting the secretion of anti-inflammatory cytokines, such as TGFβ and interleukin-10 (IL-10) [39]. Macrophages recognize the apoptotic cells via surface ligands, among them phosphatidylserine (PS), and "silently" clear the cells [40, 41, 42]. In humans, apoptosis after MI occurs mainly in the border zones and in the remote areas of ischemia, which may lead to only minor resolution of inflammation [43].

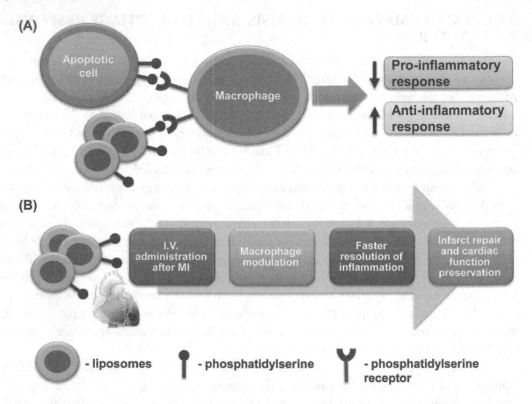

Figure 9.2: The modulation of macrophages by PS-presenting liposomes as an effective strategy for infarct repair. **A.** Nature-mimicking mechanism of macrophage modulation and phenotype switch by PS-presenting liposomes. **B.** Treatment strategy and proposed mechanism of infarct repair by PS-presenting liposomes.

Our strategy is based on the exogenous administration of PS-presenting liposomes, designed to mimic the apoptotic cells and their anti-inflammatory effects (Fig. 9.2).

At first, we demonstrated that following the uptake of PS-presenting liposomes by macrophages *in vitro*, the cells secreted high levels of anti-inflammatory cytokines, TGFβ and IL-10, and upregulated the expression of the mannose receptor CD206 (a surface marker of reparative macrophages), concomitant with down-regulation of the surface marker of pro-inflammatory macrophages, CD86. An intramyocardial injection of PS-presenting liposomes induced cardiac macrophages to secrete anti-inflammatory cytokines (TGFβ and IL-10) and reduced secretion of pro-inflammatory cytokine tumor necrosis factor-α (TNF-α) as early as three days after treatment, one day earlier than without treatment (Fig. 9.3A).

We then demonstrated the ability of PS-presenting liposomes to target the infarct after i.v. injection and be engulfed by cardiac macrophages using liposomes containing iron-oxide and their follow-up by magnetic resonance imaging (MRI) and immuno-staining of cross-sections for ED-1 (macrophage marker, brown stain) and for iron (blue stain). The liposomes were injected through the femoral vein, 48 h after MI induction in rats. MRI scans and histology of the hearts 4 d later revealed the presence of resident and/or infiltrating macrophages at the infarct that had taken up PS-presenting liposomes and accumulated in the infarct (Fig. 9.3B).

The finding that PS-presenting liposomes can be targeted to the infarct after i.v. administration and be taken up by cardiac macrophages prompted us to test the efficacy of this strategy to improve infarct repair after MI. The treatment was performed 48 h after MI induction because at this time point the macrophages are found in greater numbers at the infarct than immediately after MI. Treatment with PS-presenting liposomes by i.v. administration enhanced angiogenesis at the infarct compared to infarcts treated with PS-lacking liposomes or saline; it preserved infarct thickness to a better extent and prevented infarct expansion. Echocardiography studies showed that LV end systolic and diastolic areas (LVES and LVED areas) were more preserved after the treatment with PS-presenting liposomes, suggesting that this treatment is capable of preventing the LV remodeling associated with MI (Fig. 9.3C) [44].

Our results suggest that the modulation of recruited or resident cardiac macrophages by applying PS-presenting liposomes is feasible, leading to attenuation in left ventricle remodeling and prevention of heart dilatation. Its nature-mimicking mechanism, defined composition, and plausible delivery options make this strategy a unique and applicable approach for MI treatment. With respect to translation into the clinics, the strategy of using autologous apoptotic cells to treat chronic heart failure has been clinically tested (ACCLAIM trial) showing some benefit for the treatment [37]. Yet, there are concerns using autologous apoptotic cells because this treatment can ameliorate autoimmune diseases, for example, via the release of auto-antigens. The use of well-defined PS-presenting liposomes abrogates the concerns associated with apoptotic cell treatment, while still benefiting from their anti-inflammatory effect and positive effects on infarct repair.

9.8   SUMMARY AND CONCLUSIONS

Acellular biomaterials and scaffolds in various forms were shown to be effective for myocardial repair, by creating a more favorable environment for healing, while simultaneously providing mechanical support to the infarcted wall. The clinical applicability of the acellular biomaterial strategy has been intensified with the development and use of injectable forms of biomaterials, delivered as hydrogels

(A)

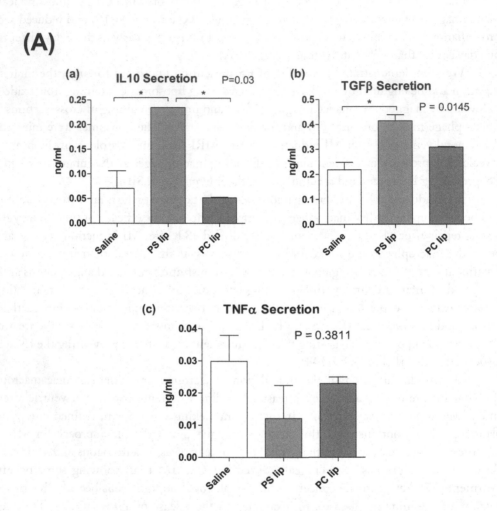

Figure 9.3: PS-presenting liposomes as an effective strategy for immunomodulation and infarct repair. **A.** Immunomodulation of cardiac macrophages after MI. MI was induced in mice, followed by intramyocardial injections of either PS-presenting liposomes (PS lip), PS-lacking liposomes (PC lip), or saline. Three days later, macrophages were isolated from the infarcted hearts, cultured for 24 h and the collected culture medium was analyzed by the respective ELISAs using antibodies against anti-inflammatory (IL-10 (**a**) and TGFβ (**b**)), and pro-inflammatory (TNFα (**c**)) cytokines. * denotes statistically significant difference.

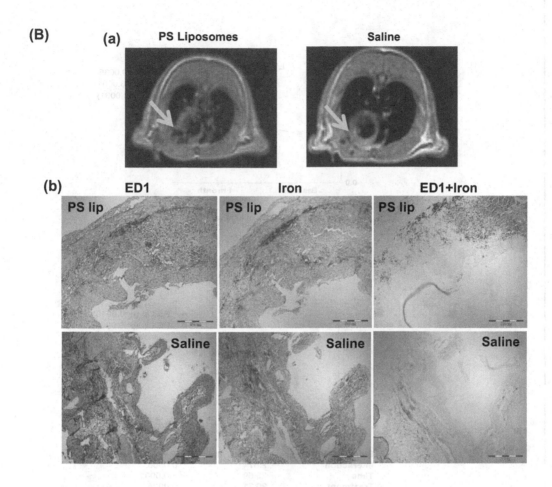

Figure 9.3: PS-presenting liposomes as an effective strategy for immunomodulation and infarct repair. **B.** *In vivo* uptake and accumulation of PS-presenting liposomes in cardiac macrophages after i.v. administration to rats, 48 hours after MI induction. **a.** MRI scans of *in vivo* uptake and accumulation of PS-presenting liposomes, 4 days after injection. The dark areas in the coronal sections represent macrophages, which have uptaken PS-presenting liposomes containing the iron oxide. **b.** Histology/immuno-histochemistry of cross-sections from mice hearts. Four days after liposome injection, the animals were sacrificed and hearts were fixated and sliced for histology and immuno-staining for ED1 (a marker for resident macrophages, brown) and iron oxide (blue). Bar = 500 μm.

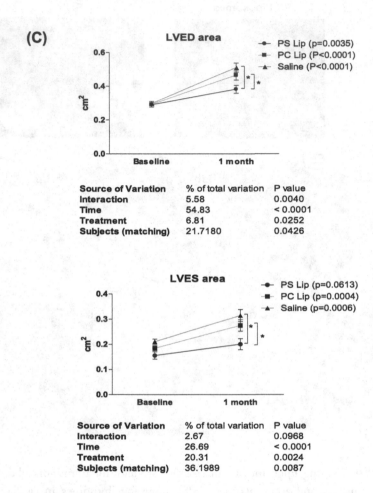

Figure 9.3: PS-presenting liposomes as an effective strategy for immunomodulation and infarct repair. C. Effect of i.v. injected PS-presenting liposomes on LV remodeling after MI. Results of echocardiography study for measured LV end systolic and diastolic (LVES and LVED) areas. * denotes statistically significant difference [44].

or solutions. The promising potential of this strategy in human therapy is exemplified by alginate biomaterial, currently being evaluated in advanced clinical trials in MI patients.

The mechanism behind the positive effects of sole biomaterial injection is assumed to be mainly passive, as it primarily affects the mechanical properties of the ventricle, by providing mechanical support and reducing wall stress. None of the studies have shown active regeneration and the addition of new contractile tissue after treatment. In the next chapter, we will describe the application of these biomaterial platforms in combination with bioactive molecules and signals to achieve this goal.

BIBLIOGRAPHY

[1] Akhyari P, Kamiya H, Haverich A, Karck M, Lichtenberg A. Myocardial tissue engineering: the extracellular matrix. Eur J Cardiothorac Surg. 2008;34:229–41. DOI: 10.1016/j.ejcts.2008.03.062 118

[2] Dobaczewski M, Gonzalez-Quesada C, Frangogiannis NG. The extracellular matrix as a modulator of the inflammatory and reparative response following myocardial infarction. J Mol Cell Cardiol. 2010;48:504–11. DOI: 10.1016/j.yjmcc.2009.07.015 118

[3] Leor J, Aboulafia-Etzion S, Dar A, Shapiro L, Barbash IM, Battler A, et al. Bioengineered cardiac grafts. A new approach to repair the infarcted myocardium? Circulation. 2000;102 (supplII):56–61. 118

[4] Leor J, Amsalem Y, Cohen S. Cells, scaffolds, and molecules for myocardial tissue engineering. Pharmacol Ther. 2005;105:151–63. DOI: 10.1093/bmb/ldn026 118

[5] Callegari A, Bollini S, Iop L, Chiavegato A, Torregrossa G, Pozzobon M, et al. Neovascularization induced by porous collagen scaffold implanted on intact and cryoinjured rat hearts. Biomaterials. 2007;28:5449–61. DOI: 10.1016/j.biomaterials.2007.07.022 118

[6] Gaballa MA, Sunkomat JN, Thai H, Morkin E, Ewy G, Goldman S. Grafting an acellular 3-dimensional collagen scaffold onto a non-transmural infarcted myocardium induces neoangiogenesis and reduces cardiac remodeling. J Heart Lung Transplant. 2006;25:946–54. DOI: 10.1016/j.healun.2006.04.008 118

[7] Fujimoto KL, Tobita K, Merryman WD, Guan J, Momoi N, Stolz DB, et al. An elastic, biodegradable cardiac patch induces contractile smooth muscle and improves cardiac remodeling and function in subacute myocardial infarction. J Am Coll Cardiol. 2007;49:2292–300. DOI: 10.1016/j.jacc.2007.02.050 118

[8] Badylak SF, Taylor D, Uygun K. Whole-organ tissue engineering: decellularization and recellularization of three-dimensional matrix scaffolds. Annu Rev Biomed Eng. 2011;13:27–53. DOI: 10.1146/annurev-bioeng-071910-124743 119

[9] Kochupura PV, Azeloglu EU, Kelly DJ, Doronin SV, Badylak SF, Krukenkamp IB, et al. Tissue-engineered myocardial patch derived from extracellular matrix provides regional mechanical function. Circulation. 2005;112:I144–9. DOI: 10.1161/CIRCULATIONAHA.104.524355 119

[10] Ott HC, Matthiesen TS, Goh SK, Black LD, Kren SM, Netoff TI, et al. Perfusion-decellularized matrix: using nature's platform to engineer a bioartificial heart. Nature medicine. 2008;14:213–21. DOI: 10.1038/nm1684 119

[11] Eitan Y, Sarig U, Dahan N, Machluf M. Acellular cardiac extracellular matrix as a scaffold for tissue engineering: In-vitro cell support, remodeling and biocompatibility. Tissue Eng Part C Methods. 2009;16:671–83. DOI: 10.1089/ten.tec.2009.0111 119

[12] Singelyn JM, Christman KL. Injectable materials for the treatment of myocardial infarction and heart failure: the promise of decellularized matrices. J Cardiovasc Transl Res. 2010;3:478–86. DOI: 10.1007/s12265-010-9202-x 119

[13] Robinson KA, Li J, Mathison M, Redkar A, Cui J, Chronos NA, et al. Extracellular matrix scaffold for cardiac repair. Circulation. 2005;112:I135–43. DOI: 10.1161/CIRCULATIONAHA.104.525436 119

[14] Wainwright JM, Czajka CA, Patel UB, Freytes DO, Tobita K, Gilbert TW, et al. Preparation of Cardiac Extracellular Matrix from an Intact Porcine Heart. Tissue Eng Part C Methods. 2009;16:525–32. DOI: 10.1089/ten.tec.2009.0392 119

[15] Mirsadraee S, Wilcox HE, Korossis SA, Kearney JN, Watterson KG, Fisher J, et al. Development and characterization of an acellular human pericardial matrix for tissue engineering. Tissue Eng. 2006;12:763–73. DOI: 10.1089/ten.2006.12.763 119

[16] Rane AA, Christman KL. Biomaterials for the treatment of myocardial infarction a 5-year update. J Am Coll Cardiol. 2011;58:2615–29. DOI: 10.1016/j.jacc.2011.11.001 120

[17] Christman KL, Lee RJ. Biomaterials for the treatment of myocardial infarction. J Am Coll Cardiol. 2006;48:907–13. DOI: 10.1016/j.jacc.2006.06.005 120

[18] Tous E, Purcell B, Ifkovits JL, Burdick JA. Injectable acellular hydrogels for cardiac repair. J Cardiovasc Transl Res. 2011;4:528–42. DOI: 10.1007/s12265-011-9291-1 120, 121

[19] Christman KL, Fok HH, Sievers RE, Fang Q, Lee RJ. Fibrin glue alone and skeletal myoblasts in a fibrin scaffold preserve cardiac function after myocardial infarction. Tissue Eng. 2004;10:403–9. DOI: 10.1089/107632704323061762 120

[20] Christman KL, Vardanian AJ, Fang Q, Sievers RE, Fok HH, Lee RJ. Injectable fibrin scaffold improves cell transplant survival, reduces infarct expansion, and induces neovasculature formation in ischemic myocardium. J Am Coll Cardiol. 2004;44:654–60. DOI: 10.1016/j.jacc.2004.04.040 120

[21] Mukherjee R, Zavadzkas JA, Saunders SM, McLean JE, Jeffords LB, Beck C, et al. Targeted myocardial microinjections of a biocomposite material reduces infarct expansion in pigs. Ann Thorac Surg. 2008;86:1268–76. DOI: 10.1016/j.athoracsur.2008.04.107 120

[22] Lu WN, Lu SH, Wang HB, Li DX, Duan CM, Liu ZQ, et al. Functional improvement of infarcted heart by co-injection of embryonic stem cells with temperature-responsive chitosan hydrogel. Tissue Eng Part A. 2009;15:1437–47. DOI: 10.1089/ten.tea.2008.0143 121

[23] Dai W, Wold LE, Dow JS, Kloner RA. Thickening of the infarcted wall by collagen injection improves left ventricular function in rats: a novel approach to preserve cardiac function after myocardial infarction. J Am Coll Cardiol. 2005;46:714–9. DOI: 10.1016/j.jacc.2005.04.056 121

[24] Ifkovits JL, Tous E, Minakawa M, Morita M, Robb JD, Koomalsingh KJ, et al. Injectable hydrogel properties influence infarct expansion and extent of postinfarction left ventricular remodeling in an ovine model. Proceedings of the National Academy of Sciences of the United States of America. 2010;107:11507–12. DOI: 10.1073/pnas.1004097107 121

[25] Wang T, Wu DQ, Jiang XJ, Zhang XZ, Li XY, Zhang JF, et al. Novel thermosensitive hydrogel injection inhibits post-infarct ventricle remodelling. Eur J Heart Fail. 2009;11:14–9. DOI: 10.1093/eurjhf/hfn009 122

[26] Seif-Naraghi SB, Salvatore MA, Schup-Magoffin PJ, Hu DP, Christman KL. Design and characterization of an injectable pericardial matrix gel: a potentially autologous scaffold for cardiac tissue engineering. Tissue Eng Part A. 2010;16:2017–27. DOI: 10.1089/ten.tea.2009.0768 122

[27] Singelyn JM, DeQuach JA, Seif-Naraghi SB, Littlefield RB, Schup-Magoffin PJ, Christman KL. Naturally derived myocardial matrix as an injectable scaffold for cardiac tissue engineering. Biomaterials. 2009;30:5409–16. DOI: 10.1016/j.biomaterials.2009.06.045 122

[28] Singelyn JM, Sundaramurthy P, Johnson TD, Schup-Magoffin PJ, Hu DP, Faulk DM, et al. Catheter-deliverable hydrogel derived from decellularized ventricular extracellular matrix increases endogenous cardiomyocytes and preserves cardiac function post-myocardial infarction. J Am Coll Cardiol. 2012;59:751–63. DOI: 10.1016/j.jacc.2011.10.888 122, 123

[29] Landa N, Miller L, Feinberg MS, Holbova R, Shachar M, Freeman I, et al. Effect of injectable alginate implant on cardiac remodeling and function after recent and old infarcts in rat. Circulation. 2008;117:1388–96. DOI: 10.1161/CIRCULATIONAHA.107.727420 123, 124, 125, 126, 127, 128, 129, 130

[30] Leor J, Tuvia S, Guetta V, Manczur F, Castel D, Willenz U, et al. Intracoronary injection of in situ forming alginate hydrogel reverses left ventricular remodeling after myocardial infarction

in Swine. J Am Coll Cardiol. 2009;54:1014–23. DOI: 10.1016/j.jacc.2009.06.010 123, 124, 125, 126, 127, 128, 129, 130

[31] Tsur-Gang O, Ruvinov E, Landa N, Holbova R, Feinberg MS, Leor J, et al. The effects of peptide-based modification of alginate on left ventricular remodeling and function after myocardial infarction. Biomaterials. 2009;30:189–95. DOI: 10.1016/j.biomaterials.2008.09.018 123

[32] BioLineRx L. Safety and Feasibility of the Injectable BL-1040 Implant. Study NCT00557531, 2009. Available at: http://www.ClinicalTrials.gov. Accessed December 27, 2011. 125

[33] Wall ST, Walker JC, Healy KE, Ratcliffe MB, Guccione JM. Theoretical impact of the injection of material into the myocardium: a finite element model simulation. Circulation. 2006;114:2627–35. DOI: 10.1161/CIRCULATIONAHA.106.657270 130

[34] Lambert JM, Lopez EF, Lindsey ML. Macrophage roles following myocardial infarction. Int J Cardiol. 2008;130:147–58. DOI: 10.1016/j.ijcard.2008.04.059 131

[35] Nahrendorf M, Swirski FK, Aikawa E, Stangenberg L, Wurdinger T, Figueiredo JL, et al. The healing myocardium sequentially mobilizes two monocyte subsets with divergent and complementary functions. J Exp Med. 2007;204:3037–47. DOI: 10.1084/jem.20070885 131

[36] Troidl C, Mollmann H, Nef H, Masseli F, Voss S, Szardien S, et al. Classically and alternatively activated macrophages contribute to tissue remodelling after myocardial infarction. J Cell Mol Med. 2009;13:3485–96. DOI: 10.1111/j.1582-4934.2009.00707.x 131

[37] Torre-Amione G, Anker SD, Bourge RC, Colucci WS, Greenberg BH, Hildebrandt P, et al. Results of a non-specific immunomodulation therapy in chronic heart failure (ACCLAIM trial): a placebo-controlled randomised trial. Lancet. 2008;371:228–36. DOI: 10.1016/S0140-6736(08)60914-9 131, 133

[38] Leor J, Rozen L, Zuloff-Shani A, Feinberg MS, Amsalem Y, Barbash IM, et al. Ex vivo activated human macrophages improve healing, remodeling, and function of the infarcted heart. Circulation. 2006;114:I94–100. DOI: 10.1161/CIRCULATIONAHA.105.000331 131

[39] Fürnrohr BG, Sheriff A, Munoz L, von Briesen H, Urbonaviciute V, Neubert K, et al. Signals, receptors, and cytokines involved in the immunomodulatory and anti-inflammatory properties of apoptotic cells. Signal Transduction. 2005;5:356–65. DOI: 10.1002/sita.200500071 132

[40] Bose J, Gruber AD, Helming L, Schiebe S, Wegener I, Hafner M, et al. The phosphatidylserine receptor has essential functions during embryogenesis but not in apoptotic cell removal. J Biol. 2004;3:15. DOI: 10.1186/jbiol10 132

[41] Fadok VA, Voelker DR, Campbell PA, Cohen JJ, Bratton DL, Henson PM. Exposure of phosphatidylserine on the surface of apoptotic lymphocytes triggers specific recognition and removal by macrophages. J Immunol. 1992;148:2207–16. 132

[42] Huynh ML, Fadok VA, Henson PM. Phosphatidylserine-dependent ingestion of apoptotic cells promotes TGF-beta1 secretion and the resolution of inflammation. J Clin Invest. 2002;109:41–50. DOI: 10.1172/JCI11638 132

[43] Thum T, Bauersachs J, Poole-Wilson PA, Volk HD, Anker SD. The dying stem cell hypothesis: immune modulation as a novel mechanism for progenitor cell therapy in cardiac muscle. J Am Coll Cardiol. 2005;46:1799–802. DOI: 10.1016/j.jacc.2005.07.053 132

[44] Harel-Adar T, Ben Mordechai T, Amsalem Y, Feinberg MS, Leor J, Cohen S. Modulation of cardiac macrophages by phosphatidylserine-presenting liposomes improves infarct repair. Proceedings of the National Academy of Sciences of the United States of America. 2011;108:1827–32. DOI: 10.1073/pnas.1015623108 133, 136

CHAPTER 10

Biomaterial-based Controlled Delivery of Bioactive Molecules for Myocardial Regeneration

CHAPTER SUMMARY

As the therapeutic benefits of biomaterial treatments of MI and the paracrine effects of stem cells on cardiac regeneration are being established, a new strategy has emerged combining both bioactive molecules and biomaterials to achieve effective therapy. In this strategy, the biomaterials function both as supporting and ECM replacing platforms as well as local depots for controlled biomolecule delivery, in aim to achieve a long-term active form of myocardial regeneration. The present chapter presents the principles of this strategy, the type of growth factors and cytokines found to be inducers of myocardial regeneration, and the various strategies for their incorporation in different biomaterial systems to achieve protection and sustained presentation in the infarct zone. An emphasis is given in this chapter to the features of a novel affinity-binding alginate biomaterial, whose development by our group was bio-inspired by ECM interactions with heparin-binding proteins. This platform has shown its capability for prolonged presentation of multiple growth factors and was proven to elicit beneficial therapeutic effects in several ischemic disease models.

10.1 INTRODUCTION

The data collected from various preclinical and clinical trials clearly show that stem cell transplantation, although having some initial beneficial effects, have failed to provide long-term improvements in cardiac function. This has been mainly attributed to the low cell engraftment at the infarct and the lack of a true regeneration, i.e., differentiation of the transplanted cells into cardiomyocytes. The initial beneficial effects observed in some of these studies were mainly ascribed to the action of various soluble cell-secreted bioactive molecules, such as growth factors and cytokines. Indeed, studies that examined the effect of systemic delivery of exogenous bioactive molecules revealed positive effects on cardiac function, however, they also presented several drawbacks, such as the requirement for repeated and high doses, possible mutagenesis (when using viral vectors), and safety concerns in patients, originated from the pleiotropic actions of most of the growth factors/cytokines on various organism systems.

The success of acellular biomaterials in providing *in situ* tissue support and their known property, at least of some of them, to act as depot for biological molecules, motivated the recent application of combinations of biomaterials and growth factors for cardiac repair. While the effects of sole biomaterial are mainly passive and mechanistic in nature, the combination with various bioactive molecules has the potential of adding an active component to this system, to achieve sustained and long-term effects. Both implantable constructs (scaffolds and sheets) and injectable forms (hydrogels and solutions) have been investigated as systems for bioactive molecule delivery, providing additional evidence to the versatility of biomaterial use for myocardial regeneration and repair.

10.2   EVOLUTION OF BIOACTIVE MATERIAL APPROACH FOR MYOCARDIAL REGENERATION

To achieve a long-term cardiac function restoration and/or diminish adverse LV remodeling, the therapeutic strategy should be able to induce active myocardial regeneration, e.g., to introduce viable beating tissue (Fig. 10.1) [1].

This goal can be achieved by inducing resident myocyte proliferation, migration, and activation of resident stem/progenitor cells and/or effective salvaging of existing viable functional tissue after initial infarct. Various cytokines, growth factors, and other bioactive molecules could contribute significantly to these desired effects. Yet, to maximize the efficiency of this bioactive approach, the combination of bioactive molecules with biomaterials seems to be a very attractive option. In such a way, the biomaterial will provide structural temporary matrix support and direct the formation of functional tissue, by mainly passive mechanisms mentioned before. Simultaneously, it will provide a temporary depot for sustained delivery of bioactive molecules with spatial and controlled distribution of the desired agent to induce regenerative processes [2, 3].

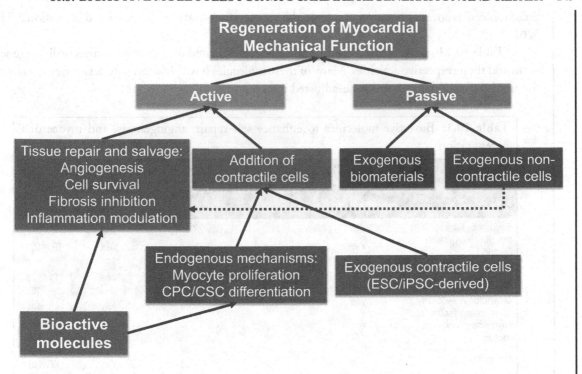

Figure 10.1: Regeneration of mechanical function in the heart can be from a passive source, such as a change in the material properties of infarcted myocardium, or from an active source. Multiple potential sources are available to improve passive function in the heart, including noncontractile cell transplantation and/or injection of biomaterials. The addition of contractile cells, however, will improve the active function of the heart and will result in long-term beneficial effects. Endogenous mechanisms for active regeneration include myocyte proliferation or endogenous CSC/CPC differentiation. A potential exogenous source of contractile cells could be ESC or iPSC-derived cardiomyocytes. In addition, various reparative processes could also contribute to tissue salvage, thus positively affecting active regeneration. Transplantation of non-contractile cells can also contribute to these processes by paracrine mechanisms (dashed line). Bioactive molecules have a broad spectrum of activities, and can induce endogenous regeneration and tissue repair mechanisms. Thus, combination of bioactive molecules with biomaterials has a promising potential in inducing active and passive regeneration simultaneously. Adapted with permission from [1].

10.3 BIOACTIVE MOLECULES FOR MYOCARDIAL REGENERATION AND REPAIR

The use of bioactive molecules (growth factors, cytokines, and stem cell mobilizing factors) is of a continuous interest in the field of therapeutic myocardial regeneration. The variable effects exerted by these molecules cover almost every target in the regeneration strategies, as discussed in Section 2.7 [4, 5, 6].

Table 10.1 lists the major bioactive molecules investigated for therapeutic myocardial regeneration and their respective activities. Many of these molecules have pleiotropic functions, emphasizing the need for careful, local, and time-adjusted interventions.

Table 10.1: Bioactive molecules to enhance self-repair, angiogenesis, and myocardial regeneration

Factor	Stem cell recruitment or mobilization	Myogenesis	Angiogenesis	Anti-apoptosis	Ref
Erythropoietin	Yes	No	Yes	Yes	[7, 8]
Insulin-like growth factor-1	Yes	Yes	Yes	Yes	[9, 10]
Fibroblast growth factor	Yes	Yes	Yes	Yes	[11, 12]
Granulocyte–colony stimulating factor	Yes	No	Yes	No	[13, 14]
Hepatocyte growth factor	Yes	Yes	Yes	Yes	[15, 16]
Periostin	No	Yes	Yes	Yes	[17, 18]
Platelet-derived growth factor-BB	Yes	No	Yes	Yes	[19, 20]
Stromal cell-derived growth factor	Yes	No	Yes	Yes	[21, 22]
Thymosin-β4	Yes	Yes	Yes	Yes	[23, 24]
Vascular endothelial growth factor	Yes	Yes	Yes	Yes	[25, 26]

The systemic delivery of some of the above growth factors was found to be beneficial for the restoration of cardiac function in animal models. However, data emerging from clinical studies has been less conclusive, as for example the case of using granulocyte colony-stimulating factor (G-CSF) [27, 28]. The mixed results obtained with systemic cytokine or growth factor administration are also accompanied by numerous safety concerns and side effects. These include an increased incidence of restenosis, elevated blood pressure and viscosity, thrombolytic events, arrhythmogenesis, and other potential detrimental effects [4, 29, 30]. In addition, systemic administration requires higher doses of the drug due to unspecific delivery, fast elimination, and extremely low protein stability in the blood. Thus, significant efforts are being invested in the development of strategies for achieving effective local and temporary delivery of various bioactive molecules by employing biomaterial-based polymeric delivery systems.

10.4 SCAFFOLD- AND HYDROGEL SHEET-BASED MOLECULE DELIVERY

Various scaffolds or sheet-like structures have been used for growth factor delivery to infarcted myocardium or for the repair of cardiac defects. Ota *et al* used a decellularized porcine urinary bladder to create a scaffold that was used for the repair of a surgically created defect in the right ventricular wall. The scaffold was loaded with fibronectin collagen-binding domain (CBD)-HGF fusion protein. The presence of CBD significantly improved HGF retention in the scaffold, probably due to its specific interactions with the collagen scaffold. The implantation of this scaffold increased contractility and electrical activity of the heart, and was associated with a homogenous repopulation by host cells and increased angiogenesis in the graft [31].

Growth factor-containing hydrogel sheets, prepared by impregnation of freeze-dried hydrogel sheets in growth factor-containing solution, have been extensively used in cardiovascular tissue engineering [32, 33]. Takehara *et al* evaluated the effect of controlled delivery of bFGF from gelatin hydrogel sheets in a chronic myocardial infarction model in pigs [34]. At four weeks after implantation, the local sustained delivery of bFGF stimulated myocardial perfusion and increased left ventricular EF. However, these effects were not compared to empty hydrogel controls. Fujiwara and co-workers used the same concept of gelatin hydrogel sheets for erythropoietin (EPO) delivery for infarct repair in rabbits. The patch was placed on the surface risk area immediately after infarct induction. Two months later, the EPO-containing hydrogel sheet improved cardiac function and reduced infarct size, compared to empty sheets or systemic EPO administration [35].

Zhang and co-workers created an SDF-1α-containing PEGylated fibrin patch by incubation of SDF-1 with PEGylated fibrinogen, followed by gelation with thrombin, and tested the effect of this delivery system in murine MI model. The patch increased c-kit$^+$ stem cell homing to myocardium and improved cardiac function [36].

10.5 INJECTABLE SYSTEMS

Injectable systems composed of biomaterial-bioactive molecule combinations offer an additional advantage to above mentioned aspects of this strategy, i.e., greater applicability in clinical settings due to the less invasive route of delivery.

Gelatin hydrogel microspheres are widely used as a carrier for growth factor delivery in various settings, including into the infarcted heart. For instance, Iwakura *et al* used gelatin hydrogel microspheres for controlled delivery of bFGF into infarcted myocardium of rats. The treatment resulted in increased angiogenesis and improved systolic and diastolic function [37, 38]. In a similar approach, Deng and coworkers injected bFGF-containing gelatin hydrogel microspheres into infarcted hearts of dogs. MRI showed improved LV function and angiogenesis [39].

Injectable self-assembling peptide nanofibers (NF) were developed by Richard Lee and colleagues and were investigated as a supporting matrix and for protein delivery into an infarcted heart [40].

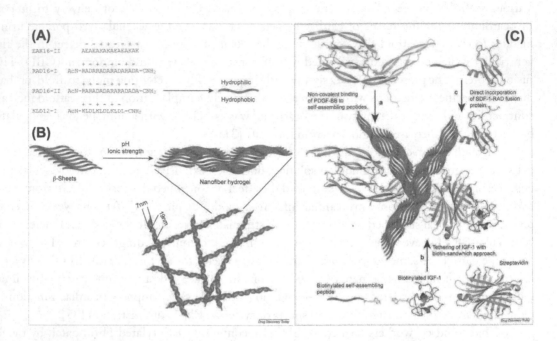

Figure 10.2: Self-assembling peptide nanofiber system for growth factor delivery. **A.** Examples of different sequences of self-assembling peptides. Hydrophobic amino acids (alanine (A) or leucine (L)) alternate with positively (lysine (K) and arginine (R)) and negatively charged amino acids (glutamate (E) and aspartate (D)). Hydrophobic amino acids are directed to one side and hydrophilic (positively and negative charged) amino acids to the other side of peptides. **B.** Self-assembling peptides are arranged in stable β-sheets at low pH and low ionic strength. Upon exposure to physiological pH and ionic strength, they form stable nanofibers with a diameter of $\sim$10 and 50–200 nm pores. **C.** Self-assembling peptides allow delivery and tethering of proteins in different manners. **(a)** Self-assembling peptides allow long-term delivery *in vivo* of PDGF-BB. PDGF-BB binds to self-assembling peptides non-covalently by electrostatic interactions. **(b)** Biotinylated variants of self-assembling peptides have been synthesized. They allow binding of every protein that can be biotinylated using streptavidin as a linker. **(c)** The sequence of self-assembling peptides can be incorporated at the N- or C-terminus of recombinant proteins, allowing incorporation of proteins in the hydrogel. Reprinted with permission from [40, 41].

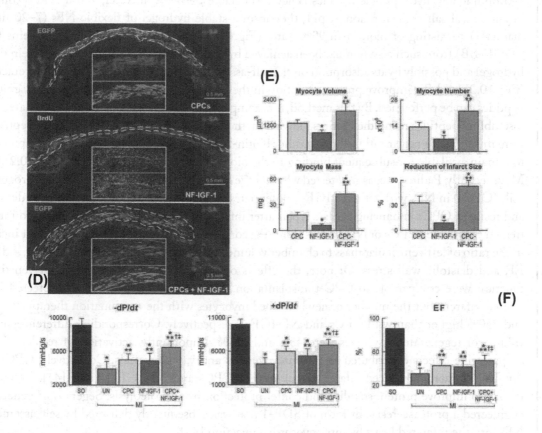

Figure 10.2: Self-assembling peptide nanofiber system for growth factor delivery. **D-F.** Myocardial repair and regeneration by combined cardiac progenitor cells (CPCs) and NF-IGF-1 therapy. **D.** Regions of regenerated myocardium (dashed line) in infarcts treated with CPCs (top panel), NF-IGF-1 (middle panel), and CPCs-NF-IGF-1 (bottom panel). In the insert, new myocytes (α-sarcomeric actin [α-SA], red) express EGFP (top and bottom panels, green) or are labeled by BrdU (middle panel, white). **E.** Average volume, number, and aggregate mass of regenerated myocytes. The latter was employed to compute infarct size. *$P<0.05$ vs. infarcts treated with CPCs only; **$P<0.05$ vs. infarcts treated with NF-IGF-1 only. **F.** Combination therapy (CPC-NF-IGF-1) improved cardiac performance. *$P<0.05$ vs. sham-operated (SO); **$P<0.05$ vs. untreated infarcts (UN); †$P<0.05$ vs. infarcts treated with CPCs; ‡$P<0.05$ vs. infarcts treated with NF-IGF-1. EF, ejection fraction; MI, myocardial infarction; SO, sham-operated; and UN, untreated infarcts. Reprinted with permission from [40, 41].

Self-assembling peptides are typically 8–16 amino acids long and are composed of alternating hydrophilic and hydrophobic residues. They form stable β-sheets in water, and upon exposure to physiological salt concentration or pH, they form a stable hydrogel of flexible NFs (7–20 nm in diameter) consisting of more than 99% water (Fig. 10.2) [40]. Slow release of the proteins (e.g., PDGF-BB) from such a system has been achieved by the physical entrapment of the protein in the hydrogel and possibly by its adsorption on the self-assembling peptides by non-covalent interactions (Fig. 10.2) [42]. To improve protein retention in the hydrogel, biotinylation of the self-assembling peptides can be performed. By this method, for example, IGF-1 was tethered to the biotinylated self-assembling peptides using the biotin sandwich method, where biotinylated IGF-1 and streptavidin were mixed in 1:1 molar ratio, allowing other biotin-binding sites on most tetravalent streptavidins to remain available for subsequent binding to the biotinylated self-assembled NFs (Fig. 10.2) [43]. More recently, Padin-Iruegas *et al* tested whether the local injection of clonogenic cardiac progenitor cells (CPCs) in NFs with tethered IGF-1 potentiates the activation and differentiation of delivered and resident CPCs enhancing cardiac repair after infarction (Fig. 10.2) [41]. Compared to infarcts treated with either CPCs or NF-IGF-1 alone, the combination therapy resulted in a greater increase in the ratio of left ventricular mass to chamber volume and a better preservation of +dP/dt, −dP/dt, EF, and diastolic wall stress. Of note, the effects of NF-IGF-1 therapy in terms of ventricular function were comparable to CPC transplantation. Myocardial regeneration was detected in all treated infarcts, but the number of newly formed myocytes with the combination therapy was 32% and 230% higher than with CPCs and NF-IGF-1, respectively. Corresponding differences in the volume of regenerated myocytes were 48% and 115%. Importantly, activation of resident CPCs by paracrine effects contributed to cardiomyogenesis and vasculogenesis. Collectively, CPCs and NF-IGF-1 therapy reduced infarct size more than CPCs and NF-IGF-1 alone [41]. In another report, to improve protein stability and confer protection from proteolysis, Segers *et al* genetically engineered a protease-resistant form of SDF-1 that was subsequently delivered by self-assembling NFs into the infarcted heart by intramyocardial injection [44].

Cardiac regeneration, most likely, would require the presentation and participation of multiple growth factors. Multiple and/or sequential factor delivery has been described by several groups, applying different delivery systems or different combination of polymers [45, 46]. Hao *et al* used a combination of partially oxidized alginates with low and high molecular weights to produce a hydrogel system that can sequentially deliver VEGF and PDGF-BB into the infarcted myocardium [47]. The factors were adsorbed to the hydrogel via electrostatic interaction and the sequential factor delivery was achieved due to the different degradation rates of the partially oxidized alginates constituting the hydrogel. One week after MI was induced in rats, the modified alginate hydrogels with loaded VEGF and PDGF-BB, were injected intramyocardially along the border of the MI. The sequential growth factor release led to a higher density of α-SMA-positive (mature) vessels compared to the delivery of single factors, and improved cardiac function.

10.6 AFFINITY-BINDING ALGINATE BIOMATERIAL FOR MULTIPLE GROWTH FACTOR DELIVERY

Bio-inspired by ECM interactions with heparin-binding proteins, our group has developed an affinity-binding alginate biomaterial to enable precise control over factor release and to allow the release of combinations of growth factors.

10.6.1 SULFATION OF ALGINATE HYDROGELS AND ANALYSIS OF BINDING

Alginate biomaterial with affinity binding sites for heparin-binding proteins was synthesized by sulfation of the uronic acid monomers in alginate, using carbodiimide chemistry (Fig. 10.3) [48].

The infrared (IR) spectrum of the product alginate-sulfate confirmed the appearance of a new major peak at 1250 cm^{-1} (assigned to S=O symmetric stretching) and a minor peak at 800 cm^{-1} (assigned to S–O–C stretching). According to nuclear magnetic resonance spectroscopy (C^{13}-NMR) spectra, the sulfate groups are added to either C-2 or C-3 or both, in an identical manner. The percentage sulfation by the Sheniger method was 8% (wt. sulfur per wt. alginate).

Surface Plasmon Resonance (SPR) analysis revealed the specific and strong binding of various heparin-binding proteins to alginate-sulfate, with equilibrium binding constants at the same order of magnitude as their binding to heparin [48, 49] (Fig. 10.3D and Table 10.2). No such interactions were recorded with pristine alginate. Thus, it appears that the binding to alginate-sulfate mimics in large the interactions of growth factors, chemokines, and cell adhesion molecules, collectively known as heparin-binding proteins (Fig. 10.3E). These molecules bind the proteoglycans heparin and heparan sulfate via high affinity, specific electrostatic interactions with the low- and high-sulfated sequences in these glycosaminoglycans (GAG) [50]. In this aspect, heparan sulfate GAGs play an important role in sequestering and storage of the proteins, and also participate in the formation of active signaling complex with a respective cell surface receptor.

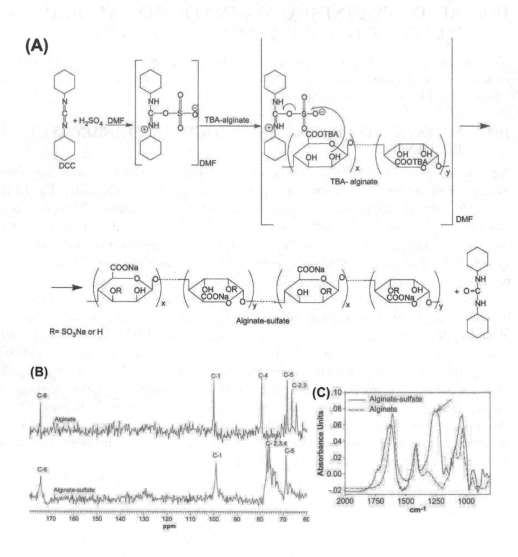

Figure 10.3: Alginate sulfation for attaining affinity-binding biomaterial. **A.** O-sulfation of the uronic acid on alginate. Reaction scheme of the sulfation of uronic acids in alginate involves the formation of protonated DCC–H_2SO_4 intermediate, followed by a hydroxyl nucleophilic attack to produce sulfated alginate and dicyclohexylurea. The latter is removed by extensive dialysis. **B.** ^{13}C NMR spectra of alginate-sulfate and raw sodium alginate, showing that sulfation occurs on C2 and C3. **C.** FTIR spectra of alginate-sulfate and raw sodium alginate. The product, alginate-sulfate, has a new major peak at $\sim$1250 cm^{-1}, assigned to S=O symmetric stretching (arrow). Reprinted with permission from [48].

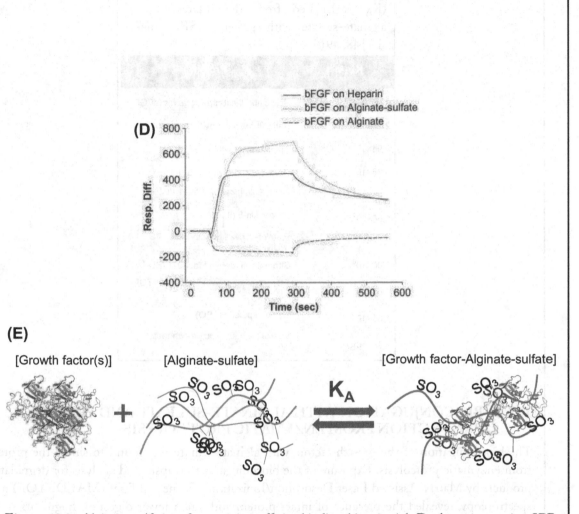

Figure 10.3: Alginate sulfation for attaining affinity-binding biomaterial. **D.** A representative SPR sensogram of bFGF binding to unmodified alginate, heparin, and alginate-sulfate, showing strong and specific binding of bFGF to heparin and alginate-sulfate, and not to unmodified alginate. **E.** The model of reversible binding. Reprinted with permission from [48].

Table 10.2: Equilibrium binding constants (K_A) calculated from the interactions of alginate-sulfate with proteins (SPR analysis) [48, 49]

K_A (M^{-1})	Protein
2.80×10^7	Acidic fibroblast growth factor (aFGF)
2.57×10^6	Basic fibroblast growth factor (bFGF)
9.93×10^6	Epidermal growth factor (EGF)
5.36×10^7	Hepatocyte growth factor (HGF)
1.01×10^8	Insulin-like growth factor-1 (IGF-1)
1.38×10^7	Interleukin-6 (IL-6)
3.53×10^7	Platelet-derived growth factor-BB (PDGF-BB)
2.06×10^8	Stromal cell-derived factor-1 (SDF-1)
6.63×10^7	Transforming growth factor- β1 (TGF-β1)
1.81×10^6	Thrombopoietin (TPO)
6.98×10^6	Vascular endothelial growth factor (VEGF)

10.6.2 BIOCONJUGATION WITH ALGINATE-SULFATE AND PROTEIN PROTECTION FROM ENZYMATIC PROTEOLYSIS

The bioconjugation of the growth factors with alginate-sulfate was found to shield the proteins from enzymatic proteolysis. Exposure of the bioconjugates to trypsin and analysis for degradation products by Matrix-Assisted Laser Desorption/Ionization – Time of Flight (MALDI-TOF) mass spectroscopy, revealed the presence of intact protein and much fewer digestion fragments in the spectrum compared to unprotected protein samples where no intact protein was detected [50, 51] (Fig. 10.4).

The shielding effect is likely due to bioconjugation leading to nanoparticle formation, wherein the alginate-sulfate found on the surface protects the core protein. The formation of nanoparticles has been established by high-resolution microscopy techniques, such as Atomic Force Microscopy (AFM), Transmission Electron Microscopy (TEM), and cryogenic-TEM, while zeta potential studies validated the presence of alginate-sulfate on the surface of the nanoparticles (Ruvinov *et al*, paper in preparation).

The effect of protein protection from proteolysis is of great importance, if the delivered proteins are to remain active for prolonged periods of time in harsh environments, where extensive protein

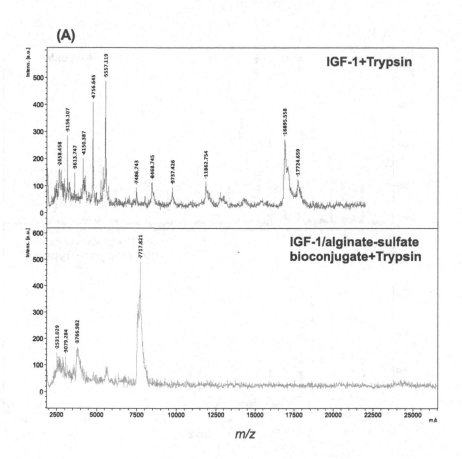

Figure 10.4: Bioconjugation with alginate-sulfate protects IGF-1 and HGF from enzymatic proteolysis. **A.** MALDI-TOF spectra of IGF-1, soluble or in the bioconjugate form. Reprinted with permission from [50, 51].

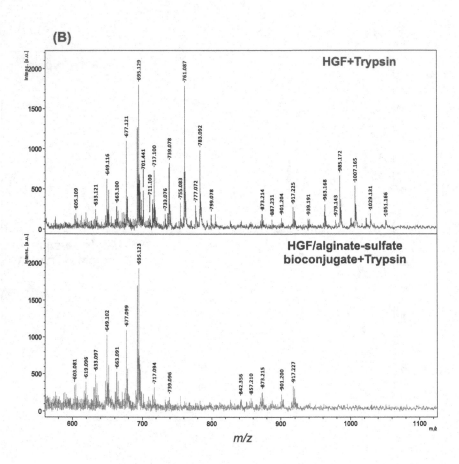

Figure 10.4: Bioconjugation with alginate-sulfate protects IGF-1 and HGF from enzymatic proteolysis. **B.** MALDI-TOF spectra of HGF, soluble or in the bioconjugate form. Reprinted with permission from [50, 51].

degradation takes place, i.e., the infarct area during the first few weeks after the initial ischemic event.

10.6.3 SCAFFOLD-BASED APPROACH USING AFFINITY-BINDING ALGINATE FOR MULTIPLE AND CONTROLLED GROWTH FACTOR DELIVERY

The combination of alginate-sulfate with pristine alginate in one device represents a unique affinity-binding alginate biomaterial, which is capable of controlling the delivery of multiple proteins, while retaining the supporting and ECM replacing properties of the alginate hydrogel.

Macroporous alginate-sulfate/alginate scaffolds were fabricated by a freeze-dry technique as described in Chapter 4 [52, 53]. The alginate-sulfate was mixed with pristine alginate solution, then the mixture was cross-linked by calcium ions and freeze-dried [49] (Fig. 10.5). As revealed by SEM analysis, incorporation of 10% (dry weight polymer) of alginate-sulfate into the alginate scaffold did not affect scaffold porosity or mechanical stability in culture (Fig. 10.5) [49].

The ability of the affinity-binding alginate scaffolds to control the release of multiple growth factors was tested using the combination of three known angiogenic factors: vascular endothelial growth factor (VEGF), platelet-derived growth factor-BB (PDGF-BB), and transforming growth factor-β1 (TGF-β1). Initial loading and binding of the factors was achieved by the addition of protein solutions to the dry scaffolds and subsequent incubation for one hour at $37°$C. *In vitro* release studies revealed a sequential order of protein release from the scaffold: VEGF was released first, followed by PDGF-BB, and then TGF-β1 [49]. Importantly, the observed release order coincided with the predicted order of the release based on the values of the equilibrium binding constants to alginate-sulfate and initial loading concentration of the factor (Fig. 10.6, also see Table 10.2). By contrast, factor release from the scaffolds lacking alginate-sulfate was rapid and was governed mainly by burst effect.

The sequential delivery of VEGF, PDGF-BB, and TGF-β1 from the affinity-binding scaffold mimics the signal cascade acting in angiogenesis, namely the initiation of the process by VEGF with endothelial cell (EC) assembly, followed by PDGF-BB-mediated smooth muscle cell and pericyte recruitment, and finally, vessel remodeling and stabilization induced by TGF-β1 (Fig. 10.6B) [54, 55]. Thus, we tested the efficacy of the system to induce the formation of stable vessels. The scaffolds were implanted subcutaneously in the dorsal area in rats, and the tissues were assessed for blood vessel number and maturation, one and three months after implantation, by immunohistochemistry for lectin (a marker of endothelial cells) and α-smooth muscle actin (SMA, a marker of smooth muscle cells) (Fig. 10.7).

Consistent with the pattern of sequential factor delivery, vessel density increase was observed at one month after implantation, while effects on vessel maturation were observed after three months, as revealed by the density of α-SMA-immunostained vessels, which increased by two-fold compared to the situation after one month. By contrast, the instantaneous delivery of the three factors from pristine alginate scaffolds resulted in two-fold lower blood vessel density, smaller sized vessels and

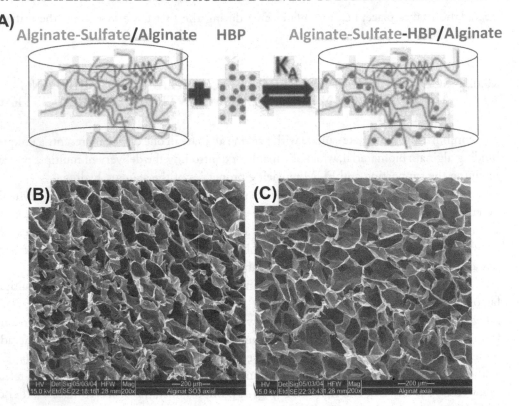

Figure 10.5: Affinity-binding alginate scaffolds for the controlled delivery of heparin-binding proteins. **A.** The concept of affinity-binding alginate scaffolds. The scaffold fabricated from alginate-sulfate/alginate can bind multiple heparin-binding proteins (HBP) via specific affinity sites on alginate-sulfate. The release rate from such scaffolds is correlated with the equilibrium binding constants (K_A) of the factors (Table 10.2). **B-C.** SEM visualization of internal morphology of affinity-binding (alginate-sulfate containing) (**B**) or unmodified alginate (**C**) scaffolds. Reprinted with permission from [49].

a similar percentage of mature vessels at one and three months, indicating the short-term effect of the adsorbed factors on scaffold vascularization [49].

In a later study, alginate scaffolds with affinity-bound mixture of pro-survival and angiogenic factors (SDF-1, IGF-1, and VEGF) were also used for the creation of a vascularized cardiac patch by pre-implantation into the omentum for 7 days (Section 8.3) [56]. Such omentum-generated pre-vascularized cardiac patch showed improved structural and electrical integration into host myocardium. Moreover, the vascularized patch induced thicker scars, prevented further dilatation of the chamber and ventricular dysfunction. Interestingly, a similar scaffold, but without seeded cardiac

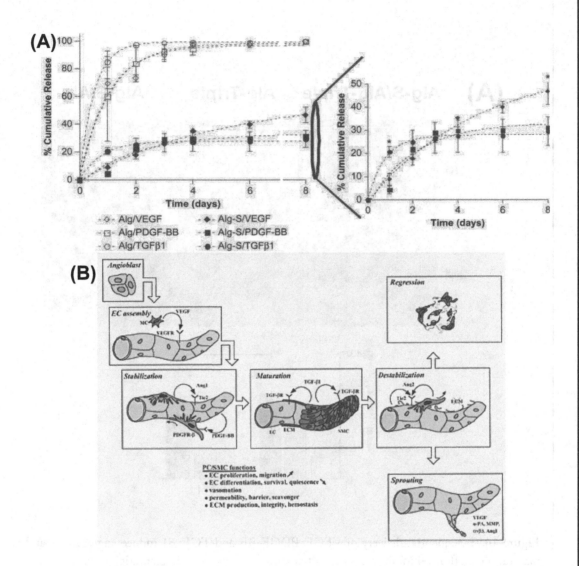

Figure 10.6: Sequential release from affinity-binding alginate scaffolds. **A.** Sequential delivery of VEGF, PDGF-BB, and TFG-β1 from alginate-sulfate/alginate scaffolds is observed, while from alginate scaffolds, the same factors are released in a burst. The right panel is a magnification of the factor release pattern from the affinity-binding alginate scaffolds. **B.** The release order (VEGF, PDGF-BB, and TGF-β1) correlates with the sequence of events during angiogenesis. Reprinted with permission from [49, 54].

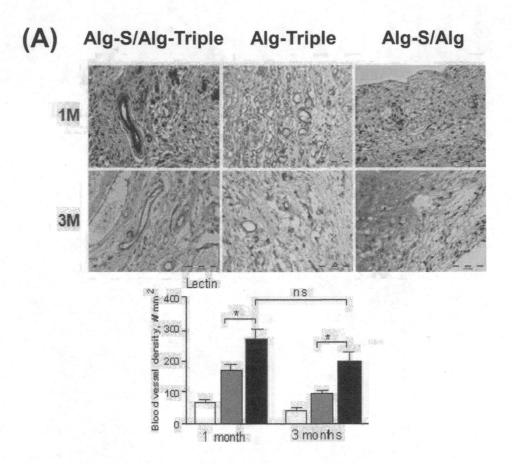

Figure 10.7: Sequential delivery of VEGF, PDGF-BB and TGF-β1 induce angiogenesis and vessel maturation in affinity-binding alginate scaffolds after subcutaneous implantation in rats. Quantification of blood vessel densities by counting (**A**) lectin-positive vessels or (**B**, *next page*) SMA-positive vessels in sections from implanted scaffolds retrieved after 1 and 3 months. Scale bar 100 μm. Reprinted with permission from [49].

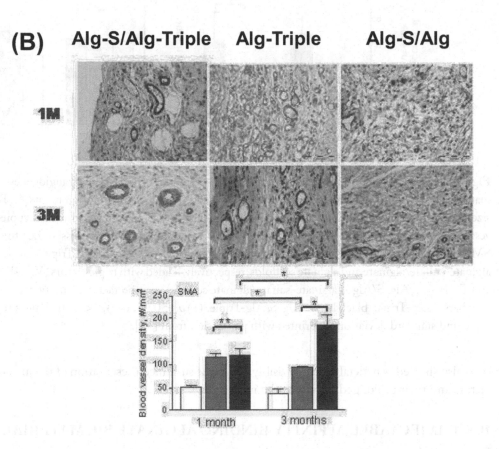

Figure 10.7: Sequential delivery of VEGF, PDGF-BB and TGF-β1 induce angiogenesis and vessel maturation in affinity-binding alginate scaffolds after subcutaneous implantation in rats. Quantification of blood vessel densities by counting (**A,** *previous page*) lectin-positive vessels or (**B**) SMA-positive vessels in sections from implanted scaffolds retrieved after 1 and 3 months. Scale bar 100 μm. Reprinted with permission from [49].

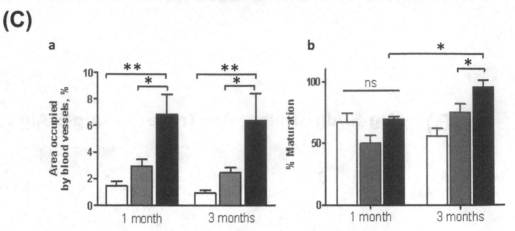

Figure 10.7: Sequential delivery of VEGF, PDGF-BB and TGF-β1 induce angiogenesis and vessel maturation in affinity-binding alginate scaffolds after subcutaneous implantation in rats. **C.** Blood vessel size and maturation in the implanted scaffolds as judged by (a) percentage of area occupied by blood vessels determined on lectin-stained sections, and (b) percentage of matured vessels, i.e., the ratio of α-SMA-positive vessel density to lectin-positive vessel density × 100. Alg-S/Alg-Triple and Alg-Triple are alginate-sulfate/alginate and alginate scaffolds, respectively, loaded with triple factors (VEGF, PDGF-BB and TGF-β1). Alg-S/Alg is alginate-sulfate/alginate scaffold with no factors. Empty bars - Alg-S/Alg; grey bars - Alg-Triple; black bars - Alg-S/Alg-Triple. (∗)$p < 0.05$, (∗∗)$p < 0.01$. Values represent the mean and standard deviation. Reprinted with permission from [49].

cells, also showed beneficial results, possibly due to the sustained presentation of the included growth factors and their prolonged activity at the infarct zone.

10.6.4   INJECTABLE AFFINITY-BINDING ALGINATE BIOMATERIAL

Features of the System

To enhance the clinical applicability of the affinity-binding alginate system for myocardial repair, it was fabricated in an injectable form. This has been achieved by mixing the alginate-sulfate/growth factor bioconjugates with partially cross-linked alginate solution, whose preparation and properties are described in Section 9.5 (Fig. 10.8). The preparation of such a system is simple and rapid (∼2 hours), compared to the far more elaborative processes described above for the fabrication of delivery systems for multiple growth factors.

The injectable affinity-binding alginate system was shown to be easily delivered into the infarcted heart where it created a hydrogel *in situ*, capable of the sustained delivery of growth factors. As such the resultant system has dual functions: 1) it may confer temporary tissue support and replace damaged ECM, together with temporary mechanical stabilization of the infarct, as previously shown

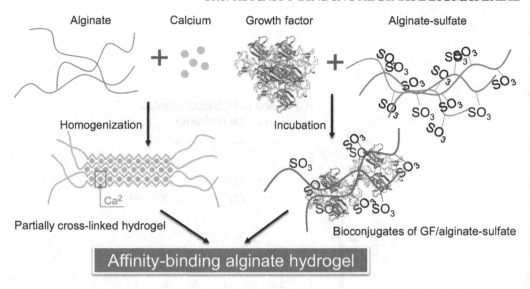

Figure 10.8: Schematic representation of the preparation methodology of injectable affinity-binding alginate system.

for this unique alginate biomaterial; 2) it should control the release and presentation of multiple growth factors. In the next section, we present several animal studies, using different disease models and different combination of growth factors, which provide proof-of-concept for the efficacy of the injectable affinity-binding alginate system.

Dual Growth Factor Release and Tissue Retention at the Infarct

Among the known available bioactive molecules, insulin-like growth factor-1 (IGF-1) and hepatocyte growth factor (HGF) are well known as potent cardiovascular-protective proteins, affecting several major aspects of myocardial regeneration. IGF-1 is a strong antiapoptotic factor in different cell types, including cardiomyocytes [10, 57, 58, 59]. Due to its marked cardioprotection effect, IGF-1 administration can improve cardiac function after MI [6]. HGF is a strong proangiogenic and antifibrotic factor [30, 60, 61, 62, 63]. In addition, HGF, together with IGF-1, induced resident cardiac stem cell migration and activation, that led to new myocardium formation in dogs [64, 65]. Due to their established and complimentary beneficial effects on infarcted myocardium, these proteins were chosen as bioactive components of the injectable delivery system. Strong affinity binding of both proteins to alginate-sulfate has been previously confirmed (Table 10.2) [48].

The release profile of the proteins from affinity-binding alginate hydrogel revealed a sequential factor delivery pattern with a greater amount of IGF-1 initially being released to the external medium until cessation on day 3, while HGF continued to be released and accumulated in the medium (Fig. 10.9) [51].

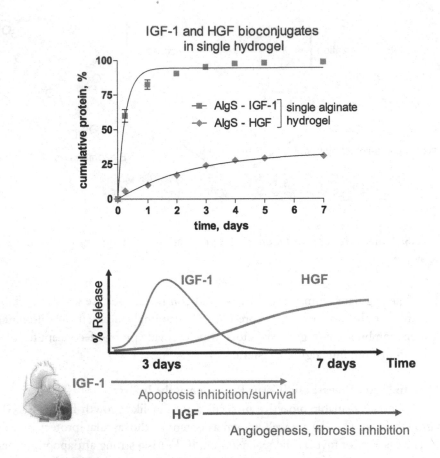

Figure 10.9: Dual IGF-1/HGF release pattern from affinity-binding alginate hydrogel. **A.** *In vitro* release from the dual factor-loaded system. **B.** The dual factor release profile coincides with the sequence of reparative processes after MI. Faster released IGF-1 could provide a strong pro-survival signal at early stages, while slower released HGF could reduce fibrosis and induce angiogenesis at later stages. Reprinted with permission from [51].

The sequential delivery of IGF-1 and HGF is suited for the proper execution of the reparative processes in the infarcted myocardium, to achieve a more favorable course of repair (Fig. 10.9). The faster released IGF-1 could provide an immediate strong pro-survival signal to rescue the functional myocardium and reduce cell apoptosis and loss after the initial ischemic event [59, 66, 67]. Processes required at later phases of repair, such as angiogenesis induction, more favorable ECM remodeling, and fibrosis reduction, can be mediated by the slower, yet continuous, release of HGF [68, 69, 70].

The released IGF-1 and HGF maintained their biological activities. Both proteins were shown to activate their respective intracellular signaling pathways (phosphorylation of AKT and ERK1/2 protein kinases for IGF-1 and HGF, respectively) and prevent cardiac cell death in an oxidative stress model [50, 51].

Increased retention of therapeutic proteins over time is one of the main attributes of a successful therapy. As mentioned, the infarct after the initial ischemic event represents a very hostile environment, where extensive protein degradation takes place as part of inflammation and ECM remodeling-induced enzymatic responses [71, 72]. Thus, we chose an acute MI and immediate post-MI injection as a model for testing the efficacy and impact of our delivery system on protein retention. HGF delivery from the affinity-binding alginate solution resulted in much greater retention and bioavailability of the factor in myocardial tissue after acute MI, as measured using anti-human HGF-specific ELISA. In contrast, soluble HGF administered by bolus injection was rapidly eliminated from the infarct [50] (Fig. 10.10).

The greater retention of HGF when delivered from the affinity-binding system is attributed to the strong, yet reversible, binding of the factor to alginate-sulfate. The *in situ* gelation of alginate forms a reservoir for the HGF/alginate-sulfate bioconjugates, thus providing an additional barrier for HGF diffusion and release.

Based on the collected data (bioconjugation and growth factor protection, sustained release and increased protein retention *in vivo*), the activity of the growth factors delivered by the *in situ* formed hydrogel with the affinity module could be the manifestation of three main processes and their combinations (Fig. 10.11): 1) the proteins are released at a rate determined by their equilibrium binding constants to alginate-sulfate and the initial loaded concentration of the individual components (affinity-binding mechanism); 2) the bioconjugates of alginate-sulfate and the factors are presented in a natural way to their respective cellular receptors, thus improving their activation and signaling, similar to their native interaction with heparan-sulfate [73]; and 3) the bioconjugates are released with time due to hydrogel dissolution, due to a decrease in calcium concentration (Fig. 10.11).

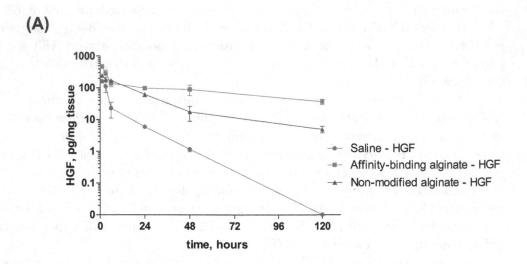

(A)

(B)

Parameter	Saline - HGF	Affinity-binding alg - HGF	Non-modified alg – HGF
AUC – measure of bioavailability, pg×h/mg tissue	1002	10322	4532

Figure 10.10: HGF retention at infarct, after injection into rat acute MI model. **A.** HGF retention profile when injected in various formulations. **B.** Nonlinear regression of data obtained from HGF retention studies in infarcted myocardium. AUC – area under the curve. Reprinted with permission from [50].

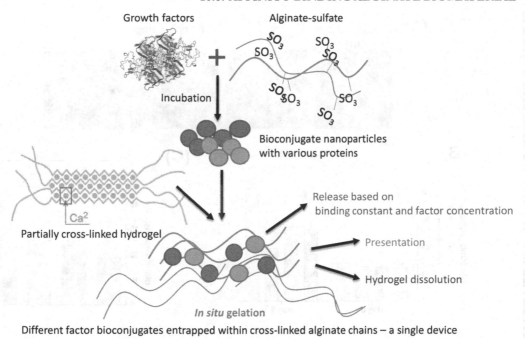

Figure 10.11: The concept of injectable affinity-binding alginate biomaterial, and the proposed modes for protein release and action.

Therapeutic Efficacy of the Combination Platform of Biomaterial and Sustained Factor Delivery
The therapeutic outcome of a single or multiple growth factor delivery by injectable affinity-binding alginate was evaluated in two ischemic disease models: 1) severe hindlimb ischemia in mice; and 2) rat model of acute MI.

Therapeutic angiogenesis in hindlimb ischemia model
Therapeutic angiogenesis is one of the key constituents for successful therapy of ischemic heart disease, peripheral artery disease, and other disorders. Induction of re-vascularization can salvage damaged ischemic tissues and facilitate self-repair [74]. As HGF is a potent angiogenic factor, we tested whether its controlled delivery by an affinity-binding alginate system would prolong and maximize its therapeutic action in a murine disease model of hindlimb ischemia [5, 70]. This strategy resulted in improved tissue perfusion, as judged by laser Doppler analysis nine days after the operation, compared to other treatment groups (Fig. 10.12).

Moreover, the treatment with HGF delivered by the affinity-binding alginate system resulted in a greater density of mature blood vessel network [50]. More recently, improved tissue blood perfusion and marked angiogenesis in the same model were observed using an injectable affinity-

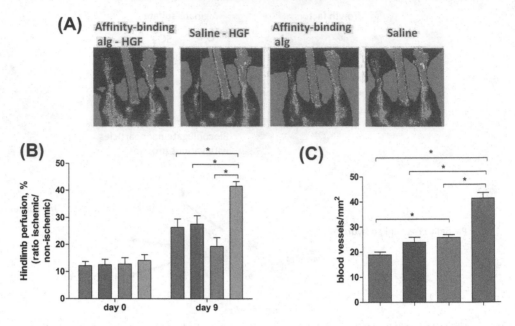

Figure 10.12: Prolonged delivery of HGF from affinity-binding alginate improves limb perfusion. **A.** Representative laser Doppler scans, 9 days after ischemia induction. **B.** Calculated perfusion percentage in different treatment groups at baseline and 9 days after operation. Green bars – saline; blue bars- affinity-binding alginate; red bars – saline-soluble HGF; brown bars – HGF-affinity-binding alginate. *P* (interaction, two-way repeated measures ANOVA) < 0.0001. $* - p < 0.05$ (Bonferroni's post-hoc test), n=10/group. **C.** Quantitative analysis of α-SMA-positive blood vessel density. *P* (one-way ANOVA) < 0.0001. Pairs indicated have significant differences ($p < 0.05$) (Bonferroni's posthoc test). n=10/group. Reprinted with permission from [50].

binding alginate biomaterial loaded with three angiogenic factors, VEGF, TGF-β1, and PDGF-BB (unpublished results).

The affinity-binding alginate was able to translate the known angiogenic effect of HGF into an improved therapeutic outcome, due to creation of a temporary favorable microenvironment for self-repair on one hand, and successful controlled delivery of the protein together with protection from enzymatic degradation and fast elimination in ischemic tissue on the other hand.

Acute MI model

Short (one week) and long (four and eight weeks) - term effects of dual (IGF-1 and HGF) factor delivery from the injectable affinity-binding alginate were tested in a rat model of acute MI. We focused on various parameters and aspects in myocardial tissue regeneration and repair, in order to examine the efficacy of the delivery system to translate known effects of the growth factors into a significant therapeutic outcome.

The sequential delivery of IGF-1 and HGF from this system reduced scar fibrosis, increased scar thickness and prevented infarct expansion, one and four weeks after injection. It also induced angiogenesis and mature blood vessel network formation, and reduced cell apoptosis at the infarct. These therapeutic effects were preserved for eight weeks after injection (unpublished data) (Fig. 10.13A).

We then evaluated the contribution of endogenous regeneration to increased tissue salvage. We focused on two processes that can be responsible for endogenous regeneration of cardiac muscle: adult cardiomyocyte proliferation (shown by staining for mitotic marker Ki-67) and the existence of cardiac stem/progenitor cells (identified by staining for transcription factor GATA-4, generally associated with cardiomyogenic differentiation) (Fig. 10.13B-D). Of note, these results were deduced based on simple morphological discrimination and single marker staining. Thus, more extensive analyzes (e.g., genetic mapping, simultaneous staining) are required to confirm the effects of sequential IGF-1/HGF delivery on endogenous myocardial regeneration.

Based on staining pattern and morphological discrimination, we identified Ki-67-positive cardiomyocytes at the infarct border of all animal groups, one week after treatment. Due to the high ultrastructural fiber organization of Ki-67-positve cells, the possible explanation for this phenomenon is cell cycle re-entry that can occur at higher incidence after MI [75]. Strikingly, the sequential delivery of IGF-1/HGF from the affinity-binding alginate biomaterial in the infarct increased the incidence of Ki-67-positive cardiomyocytes (Fig. 10.13B, D) [51]. Cell cycle re-entry could point to actual cell proliferation, an important constituent of myocardial regeneration, especially in light of recent data suggesting that endogenous myocardial regeneration could be driven by the induction of cell cycle re-entry and proliferation of existing cardiomyocytes rather than by stem or progenitor cells [76].

As an additional aspect of possible myocardial regeneration, we examined GATA-4 expression in the infarcts (Fig. 10.13C, D). Recent data show that along with its established critical role in early and late heart development and morphogenesis, GATA-4 also acts as a pro-angiogenic, anti-apoptotic, and stem cell-recruiting factor post-MI [77, 78]. The GATA-4-positive cell clusters were found only in the peri-infarct areas of animals treated with the factors, with the highest incidence detected four weeks post-MI in the animals treated with the sequentially delivered factors (Fig. 10.13D) [51]. From the cell cluster organization, we propose that these cells could be stem or progenitor cells of unknown origin. IGF-1 and HGF were previously shown to activate distinct subsets of cardiac stem cells [79]. The long-term delivery of both factors could induce the migration of these cells to the infarct region and their subsequent activation, which in turn, can partially contribute to myocardial tissue salvage. Increased GATA-4 expression could by itself contribute to pro-angiogenic and anti-apoptotic properties of the delivered proteins [77].

The extent of possible endogenous regeneration (based on Ki-67 and GATA-4 staining) after the treatment with sequentially delivered proteins in affinity-binding alginate hydrogel was

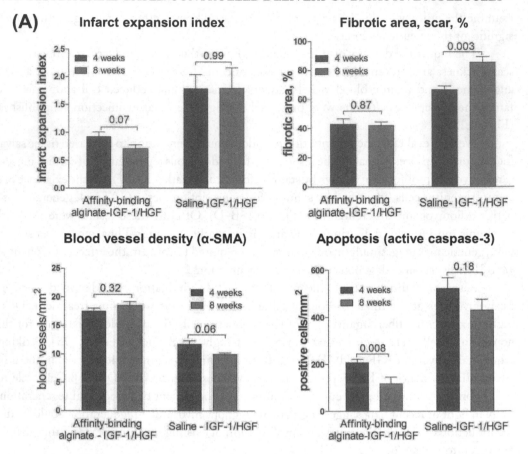

Figure 10.13: Therapeutic efficacy and cardiac tissue regeneration induced by the dual factor delivery from the injectable affinity-binding alginate system in acute MI model. **A.** The effects of a sequential delivery of IGF-1/HGF by injectable affinity-binding alginate biomaterial on various aspects of tissue regeneration. Infarct expansion index was evaluated by Masson's trichrome staining. Blood vessels were identified by α-SMA staining. Apoptotic cells were identified by active caspase-3 staining. $* - p < 0.05$. Reprinted with permission from [51] and unpublished data.

significantly reduced after a longer period of time (four weeks) [51]. This can be explained by the suboptimal concentration of the growth factors due to cessation of growth factor release and action, and also by the lack of additional endogenous signals, that could act synergistically with the delivered proteins, and are present only for a limited period of time after initial infarct. This can be overcome in future studies by administering several injections of the growth factor/biomaterial combinations.

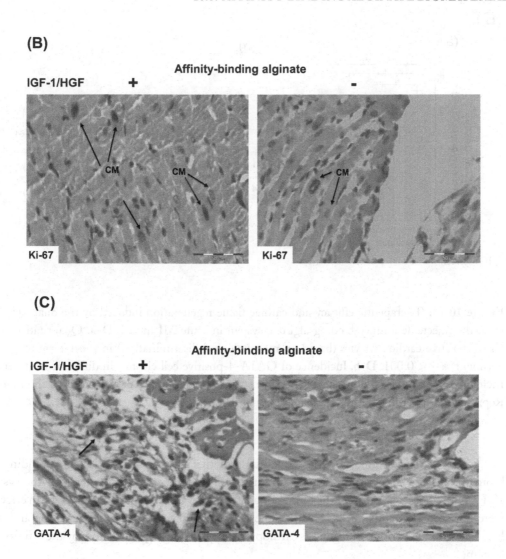

Figure 10.13: Therapeutic efficacy and cardiac tissue regeneration induced by the dual factor delivery from the injectable affinity-binding alginate system in acute MI model. **B.** Mitotic marker Ki-67 staining (brown) in healthy/infarct border region, one week after injection. Arrows show nuclear localization or unstained nuclei. Bar = 50 μm. CM-cardiomyocytes. Reprinted with permission from [51] and unpublished data. **C.** Cardiac progenitor cell marker GATA-4 staining (brown) in infarcted hearts (peri-infarct region), four weeks after injection. Arrows show nuclear localization. Bar = 50 μm. Reprinted with permission from [51] and unpublished data.

(D)

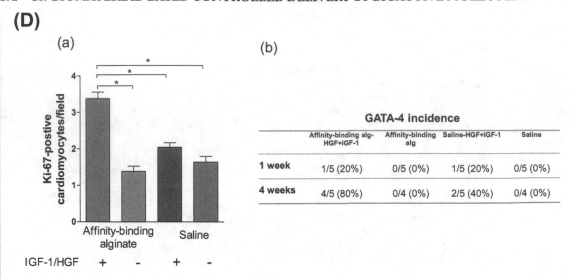

Figure 10.13: Therapeutic efficacy and cardiac tissue regeneration induced by the dual factor delivery from the injectable affinity-binding alginate system in acute MI model. **D-a.** Quantitative analysis of Ki-67-positive cardiomyocytes (based on morphological discrimination) in short-term (one week) experiment. $*- < 0.001$. **D-b.** Incidence of GATA-4-positive cell clusters in different treatment groups. Incidence = (number of animals with positive staining/total number of animals in each group) $\times$ 100. Reprinted with permission from [51] and unpublished data.

The beneficial effects of the sequential growth factor delivery by affinity-binding alginate biomaterial were shown to be specifically mediated by the active components of the system, i.e., IGF-1 and HGF. The effect of the biomaterial, if found, was limited to only short-term effects which were not seen after four weeks. This initial effect is possibly due to the sequestering and increased local retention of various endogenous cardioprotective and angiogenic factors by the affinity-binding system.

The results observed after the treatment with the sequentially delivered proteins from the affinity-binding alginate system collectively suggest that marked salvage of the functional tissue occurred due to the enhanced retention of the bioactive factors at the infarct and their sequential release pattern, mediated by the affinity-binding mechanism to alginate-sulfate and *in situ* gelation of the system (Fig. 10.11). Increased retention and prolonged release of the growth factors facilitate a more favorable course of repair, by supplying the required protective signals for longer periods of time. As a result, prolonged growth factor activity leads to more favorable remodeling at later stages (four and eight weeks after MI). These effects could also be accompanied by passive tissue support and mechanical stabilization of the infarct conferred by the alginate hydrogel itself [80, 81].

Compared to other modalities of multiple growth factor delivery described (see Section 10.5), the affinity-binding alginate system provides several potential advantages. First, the inclusion of the factors in our system does not involve covalent chemistry or non-specific binding of the factor to the biomaterial; instead it relies on mimicking the specific affinity of factors to the matrix. Second, the affinity-binding mechanism enables binding of multiple proteins and their release is well-controlled by the protein equilibrium binding constant and its initial concentration. Third, the bioconjugation between the factor and alginate-sulfate results in the formation of nanoparticles so that the proteins are shielded from enzymatic proteolysis by an external hydrogel layer [50, 51]. Finally, due to the *in situ* hydrogel formation, the affinity-binding alginate-based release system may serve as ECM replacement and provide temporal tissue support and an improved microenvironment for more effective self-repair and possible progenitor cell recruitment.

10.7   SUMMARY AND CONCLUSIONS

The prolonged delivery and activity of bioactive agents represents a major challenge in human therapy. In the cardiac settings, this mission is even more challenging as multiple factors need to be delivered in a spatio-temporal manner to induce tissue regeneration. Certain biomaterials can function both as supporting and ECM replacing platforms as well as local depots for controlled biomolecule delivery. Preferably, the biomaterials should be delivered via catheter-based approaches to enable a rapid translation into clinics. The present chapter described the various forms of biomaterial platforms aimed to deliver single or multiple bioactive molecules in a controlled manner to induce active cardiac regeneration. In particular, the focus herein was on biomaterial research performed in our lab. Bio-inspired by the affinity interactions of heparin-binding proteins and the sulfated heparin/heparan-sulfate, we developed affinity-binding alginate biomaterial by sulfation of the uronic acids on alginate. Sequential factor release patterns, designed to mimic regeneration processes, elicited therapeutic effects leading to cardiac tissue regeneration. This exciting development has yet to be investigated in long-term studies and confirmed in large animals.

BIBLIOGRAPHY

[1] Gaudette GR, Cohen IS. Cardiac regeneration: materials can improve the passive properties of myocardium, but cell therapy must do more. Circulation. 2006;114:2575–7. DOI: 10.1161/CIRCULATIONAHA.106.668707 144, 145

[2] Christman KL, Lee RJ. Biomaterials for the treatment of myocardial infarction. J Am Coll Cardiol. 2006;48:907–13. DOI: 10.1016/j.jacc.2011.11.001 144

[3] Davis ME, Hsieh PC, Grodzinsky AJ, Lee RT. Custom design of the cardiac microenvironment with biomaterials. Circ Res. 2005;97:8–15. DOI: 10.1161/01.RES.0000173376.39447.01 144

[4] Beohar N, Rapp J, Pandya S, Losordo DW. Rebuilding the damaged heart the potential of cytokines and growth factors in the treatment of ischemic heart disease. J Am Coll Cardiol. 2010;56:1287–97. DOI: 10.1016/j.jacc.2010.05.039 146

[5] Vandervelde S, van Luyn MJ, Tio RA, Harmsen MC. Signaling factors in stem cell-mediated repair of infarcted myocardium. J Mol Cell Cardiol. 2005;39:363–76. DOI: 10.1016/j.yjmcc.2005.05.012 146, 167

[6] Hausenloy DJ, Yellon DM. Cardioprotective growth factors. Cardiovasc Res. 2009;83:179–94. DOI: 10.1093/cvr/cvp062 146, 163

[7] van der Meer P, Lipsic E, Henning RH, Boddeus K, van der Velden J, Voors AA, et al. Erythropoietin induces neovascularization and improves cardiac function in rats with heart failure after myocardial infarction. J Am Coll Cardiol. 2005;46:125–33. DOI: 10.1016/j.jacc.2005.03.044

[8] Parsa CJ, Matsumoto A, Kim J, Riel RU, Pascal LS, Walton GB, et al. A novel protective effect of erythropoietin in the infarcted heart. J Clin Invest. 2003;112:999–1007. DOI: 10.1172/JCI18200

[9] Torella D, Rota M, Nurzinska D, Musso E, Monsen A, Shiraishi I, et al. Cardiac stem cell and myocyte aging, heart failure, and insulin-like growth factor-1 overexpression. Circ Res. 2004;94:514–24. DOI: 10.1161/01.RES.0000117306.10142.50

[10] Conti E, Carrozza C, Capoluongo E, Volpe M, Crea F, Zuppi C, et al. Insulin-Like Growth Factor-1 as a Vascular Protective Factor. Circulation. 2004;110:2260–5. DOI: 10.1161/01.CIR.0000144309.87183.FB 163

[11] Liao S, Porter D, Scott A, Newman G, Doetschman T, Schultz Jel J. The cardioprotective effect of the low molecular weight isoform of fibroblast growth factor-2: the role of JNK signaling. J Mol Cell Cardiol. 2007;42:106–20. DOI: 10.1016/j.yjmcc.2006.10.005

[12] Bougioukas I, Didilis V, Ypsilantis P, Giatromanolaki A, Sivridis E, Lialiaris T, et al. Intramyocardial injection of low-dose basic fibroblast growth factor or vascular endothelial growth factor induces angiogenesis in the infarcted rabbit myocardium. Cardiovasc Pathol. 2007;16:63–8. DOI: 10.1016/j.carpath.2006.08.006

[13] Harada M, Qin Y, Takano H, Minamino T, Zou Y, Toko H, et al. G-CSF prevents cardiac remodeling after myocardial infarction by activating the Jak-Stat pathway in cardiomyocytes. Nature medicine. 2005;11:305–11. DOI: 10.1038/nm1199

[14] Takano H, Ueda K, Hasegawa H, Komuro I. G-CSF therapy for acute myocardial infarction. Trends in pharmacological sciences. 2007;28:512–7. DOI: 10.1016/j.tips.2007.09.002

[15] Kondo I, Ohmori K, Oshita A, Takeuchi H, Fuke S, Shinomiya K, et al. Treatment of acute myocardial infarction by hepatocyte growth factor gene transfer: the first demonstration of myocardial transfer of a "functional" gene using ultrasonic microbubble destruction. Journal of the American College of Cardiology. 2004;44:644–53. DOI: 10.1016/j.jacc.2004.04.042

[16] Jayasankar V, Woo YJ, Bish LT, Pirolli TJ, Chatterjee S, Berry MF, et al. Gene transfer of hepatocyte growth factor attenuates postinfarction heart failure. Circulation. 2003;108 Suppl 1:II230–6. DOI: 10.1161/01.cir.0000087444.53354.66

[17] Dorn GW, 2nd. Periostin and myocardial repair, regeneration, and recovery. N Engl J Med. 2007;357:1552–4. DOI: 10.1056/NEJMcibr074816

[18] Kuhn B, del Monte F, Hajjar RJ, Chang YS, Lebeche D, Arab S, et al. Periostin induces proliferation of differentiated cardiomyocytes and promotes cardiac repair. Nature medicine. 2007;13:962–9. DOI: 10.1038/nm1619

[19] Hsieh PCH, Davis ME, Gannon J, MacGillivray C, Lee RT. Controlled delivery of PDGF-BB for myocardial protection using injectable self-assembling peptide nanofibers. The Journal of clinical investigation. 2006;116:237–48. DOI: 10.1172/JCI25878

[20] Hsieh PCH, MacGillivray C, Gannon J, Cruz FU, Lee RT. Local Controlled Intramyocardial Delivery of Platelet-Derived Growth Factor Improves Postinfarction Ventricular Function Without Pulmonary Toxicity. Circulation. 2006;114:637–44. DOI: 10.1161/CIRCULATIONAHA.106.639831

[21] Hiasa K, Ishibashi M, Ohtani K, Inoue S, Zhao Q, Kitamoto S, et al. Gene transfer of stromal cell-derived factor-1alpha enhances ischemic vasculogenesis and angiogenesis via vascular endothelial growth factor/endothelial nitric oxide synthase-related pathway: next-generation chemokine therapy for therapeutic neovascularization. Circulation. 2004;109:2454–61. DOI: 10.1161/01.CIR.0000128213.96779.61

[22] Hu X, Dai S, Wu WJ, Tan W, Zhu X, Mu J, et al. Stromal cell derived factor-1 alpha confers protection against myocardial ischemia/reperfusion injury: role of the cardiac stromal cell derived factor-1 alpha CXCR4 axis. Circulation. 2007;116:654–63. DOI: 10.1161/CIRCULATIONAHA.106.672451

[23] Bock-Marquette I, Saxena A, White MD, Michael DiMaio J, Srivastava D. Thymosin [beta]4 activates integrin-linked kinase and promotes cardiac cell migration, survival and cardiac repair. Nature. 2004;432:466–72. DOI: 10.1038/nature03000

[24] Smart N, Risebro CA, Melville AA, Moses K, Schwartz RJ, Chien KR, et al. Thymosin beta4 induces adult epicardial progenitor mobilization and neovascularization. Nature. 2007;445:177–82. DOI: 10.1038/nature05383

[25] Ferrarini M, Arsic N, Recchia FA, Zentilin L, Zacchigna S, Xu X, et al. Adeno-associated virus-mediated transduction of VEGF165 improves cardiac tissue viability and functional recovery after permanent coronary occlusion in conscious dogs. Circ Res. 2006;98:954–61. DOI: 10.1161/01.RES.0000217342.83731.89

[26] Vera Janavel G, Crottogini A, Cabeza Meckert P, Cuniberti L, Mele A, Papouchado M, et al. Plasmid-mediated VEGF gene transfer induces cardiomyogenesis and reduces myocardial infarct size in sheep. Gene therapy. 2006;13:1133–42. DOI: 10.1038/sj.gt.3302708

[27] Zohlnhofer D, Dibra A, Koppara T, de Waha A, Ripa RS, Kastrup J, et al. Stem Cell Mobilization by Granulocyte Colony-Stimulating Factor for Myocardial Recovery After Acute Myocardial Infarction: A Meta-Analysis. Journal of the American College of Cardiology. 2008;51:1429–37. DOI: 10.1016/j.jacc.2007.11.073 146

[28] Abdel-Latif A, Bolli R, Zuba-Surma EK, Tleyjeh IM, Hornung CA, Dawn B. Granulocyte colony-stimulating factor therapy for cardiac repair after acute myocardial infarction: a systematic review and meta-analysis of randomized controlled trials. Am Heart J. 2008;156:216–26. DOI: 10.1016/j.ahj.2008.03.024 146

[29] Lee TM, Chen CC, Chang NC. Granulocyte colony-stimulating factor increases sympathetic reinnervation and the arrhythmogenic response to programmed electrical stimulation after myocardial infarction in rats. American journal of physiology. 2009;297:H512–22. DOI: 10.1152/ajpheart.00077.2009 146

[30] Vandervelde S, Van Luyn, M.J.A., Tio, R.A., Harmsen, M.C. Signaling factors in stem cell-mediated repair of infarcted myocardium. Journal of molecular and cellular cardiology. 2005;39:363–76. DOI: 10.1016/j.yjmcc.2005.05.012 146, 163

[31] Ota T, Gilbert TW, Schwartzman D, McTiernan CF, Kitajima T, Ito Y, et al. A fusion protein of hepatocyte growth factor enhances reconstruction of myocardium in a cardiac patch derived from porcine urinary bladder matrix. J Thorac Cardiovasc Surg. 2008;136:1309–17. DOI: 10.1016/j.jtcvs.2008.07.008 147

[32] Masuda S, Shimizu T, Yamato M, Okano T. Cell sheet engineering for heart tissue repair. Adv Drug Deliv Rev. 2008;60:277–85. DOI: 10.1016/j.addr.2007.08.031 147

[33] Tabata Y, Ikada Y. Vascularization effect of basic fibroblast growth factor released from gelatin hydrogels with different biodegradabilities. Biomaterials. 1999;20:2169–75. DOI: 10.1016/S0142-9612(99)00121-0 147

[34] Takehara N, Tsutsumi Y, Tateishi K, Ogata T, Tanaka H, Ueyama T, et al. Controlled delivery of basic fibroblast growth factor promotes human cardiosphere-derived cell engraftment to enhance cardiac repair for chronic myocardial infarction. J Am Coll Cardiol. 2008;52:1858–65. DOI: 10.1016/j.jacc.2008.06.052 147

[35] Kobayashi H, Minatoguchi S, Yasuda S, Bao N, Kawamura I, Iwasa M, et al. Post-infarct treatment with an erythropoietin-gelatin hydrogel drug delivery system for cardiac repair. Cardiovasc Res. 2008;79:611–20. DOI: 10.1093/cvr/cvn154 147

[36] Zhang G, Nakamura Y, Wang X, Hu Q, Suggs LJ, Zhang J. Controlled release of stromal cell-derived factor-1 alpha in situ increases c-kit+ cell homing to the infarcted heart. Tissue Eng. 2007;13:2063–71. DOI: 10.1089/ten.2006.0013 147

[37] Iwakura A, Fujita M, Kataoka K, Tambara K, Sakakibara Y, Komeda M, et al. Intramyocardial sustained delivery of basic fibroblast growth factor improves angiogenesis and ventricular function in a rat infarct model. Heart Vessels. 2003;18:93–9. DOI: 10.1007/s10380-002-0686-5 147

[38] Nakajima H, Sakakibara Y, Tambara K, Iwakura A, Doi K, Marui A, et al. Therapeutic angiogenesis by the controlled release of basic fibroblast growth factor for ischemic limb and heart injury: toward safety and minimal invasiveness. J Artif Organs. 2004;7:58–61. DOI: 10.1007/s10047-004-0252-1 147

[39] Liu Y, Sun L, Huan Y, Zhao H, Deng J. Effects of basic fibroblast growth factor microspheres on angiogenesis in ischemic myocardium and cardiac function: analysis with dobutamine cardiovascular magnetic resonance tagging. Eur J Cardiothorac Surg. 2006;30:103–7. DOI: 10.1016/j.ejcts.2006.03.043 147

[40] Segers VF, Lee RT. Local delivery of proteins and the use of self-assembling peptides. Drug discovery today. 2007;12:561–8. DOI: 10.1016/j.drudis.2007.05.003 147, 148, 149, 150

[41] Padin-Iruegas ME, Misao Y, Davis ME, Segers VF, Esposito G, Tokunou T, et al. Cardiac progenitor cells and biotinylated insulin-like growth factor-1 nanofibers improve endogenous and exogenous myocardial regeneration after infarction. Circulation. 2009;120:876–87. DOI: 10.1161/CIRCULATIONAHA.109.852285 148, 149, 150

[42] Hsieh PC, Davis ME, Gannon J, MacGillivray C, Lee RT. Controlled delivery of PDGF-BB for myocardial protection using injectable self-assembling peptide nanofibers. J Clin Invest. 2006;116:237–48. DOI: 10.1172/JCI25878 150

[43] Davis ME, Hsieh PC, Takahashi T, Song Q, Zhang S, Kamm RD, et al. Local myocardial insulin-like growth factor 1 (IGF-1) delivery with biotinylated peptide nanofibers improves cell therapy for myocardial infarction. Proceedings of the National Academy of Sciences of the United States of America. 2006;103:8155–60. DOI: 10.1073/pnas.0602877103 150

[44] Segers VF, Tokunou T, Higgins LJ, MacGillivray C, Gannon J, Lee RT. Local delivery of protease-resistant stromal cell derived factor-1 for stem cell recruitment after myocardial infarction. Circulation. 2007;116:1683–92. DOI: 10.1161/CIRCULATIONAHA.107.718718 150

[45] Richardson TP, Peters MC, Ennett AB, Mooney DJ. Polymeric system for dual growth factor delivery. Nat Biotechnol. 2001;19:1029–34. DOI: 10.1038/nbt1101-1029 150

[46] Lu H, Xu X, Zhang M, Cao R, Brakenhielm E, Li C, et al. Combinatorial protein therapy of angiogenic and arteriogenic factors remarkably improves collaterogenesis and cardiac function in pigs. Proceedings of the National Academy of Sciences of the United States of America. 2007;104:12140–5. DOI: 10.1073/pnas.0704966104 150

[47] Hao X, Silva EA, Mansson-Broberg A, Grinnemo KH, Siddiqui AJ, Dellgren G, et al. Angiogenic effects of sequential release of VEGF-A(165) and PDGF-BB with alginate hydrogels after myocardial infarction. Cardiovasc Res. 2007;75:178–85.
DOI: 10.1016/j.cardiores.2007.03.028 150

[48] Freeman I, Kedem A, Cohen S. The effect of sulfation of alginate hydrogels on the specific binding and controlled release of heparin-binding proteins. Biomaterials. 2008;29:3260–8.
DOI: 10.1016/j.biomaterials.2008.04.025 151, 152, 153, 154, 163

[49] Freeman I, Cohen S. The influence of the sequential delivery of angiogenic factors from affinity-binding alginate scaffolds on vascularization. Biomaterials. 2009;30:2122–31.
DOI: 10.1016/j.biomaterials.2008.12.057 151, 154, 157, 158, 159, 160, 161, 162

[50] Ruvinov E, Leor J, Cohen S. The effects of controlled HGF delivery from an affinity-binding alginate biomaterial on angiogenesis and blood perfusion in a hindlimb ischemia model. Biomaterials. 2010;31:4573–82. DOI: 10.1016/j.biomaterials.2010.02.026 151, 154, 155, 156, 165, 166, 167, 168, 173

[51] Ruvinov E, Leor J, Cohen S. The promotion of myocardial repair by the sequential delivery of IGF-1 and HGF from an injectable alginate biomaterial in a model of acute myocardial infarction. Biomaterials. 2011;32:565–78. DOI: 10.1016/j.biomaterials.2010.08.097 154, 155, 156, 163, 164, 165, 169, 170, 171, 172, 173

[52] Shapiro L, Cohen S. Novel alginate sponges for cell culture and transplantation. Biomaterials. 1997;18:583–90. DOI: 10.1016/S0142-9612(96)00181-0 157

[53] Zmora S, Glicklis R, Cohen S. Tailoring the pore architecture in 3-D alginate scaffolds by controlling the freezing regime during fabrication. Biomaterials. 2002;23:4087–94.
DOI: 10.1016/S0142-9612(02)00146-1 157

[54] Carmeliet P. Mechanisms of angiogenesis and arteriogenesis. Nature medicine. 2000;6:389–95.
DOI: 10.1038/74651 157, 159

[55] Risau W. Mechanisms of angiogenesis. Nature. 1997;386:671–4. DOI: 10.1038/386671a0 157

[56] Dvir T, Kedem A, Ruvinov E, Levy O, Freeman I, Landa N, et al. Prevascularization of cardiac patch on the omentum improves its therapeutic outcome. Proceedings of the National Academy of Sciences of the United States of America. 2009;106:14990–5. DOI: 10.1073/pnas.0812242106 158

[57] Davani EY, Brumme Z, Singhera GK, Cote HCF, Harrigan PR, Dorscheid DR. Insulin-like growth factor-1 protects ischemic murine myocardium from ischemia/reperfusion associated injury. Critical Care. 2003;7:176–83. 163

[58] Ren J, Samson WK, Sowers JR. Insulin-like Growth FactorI as a Cardiac Hormone: Physiological and Pathophysiological Implications in Heart Disease. J Mol Cell Cardiol. 1999;31:2049–61. DOI: 10.1006/jmcc.1999.1036 163

[59] Li Q, Li B, Wang X, Leri A, Jana KP, Liu y, et al. Overexpression of Insulin-like Growth Factor-1 in Mice Protects from Myocyte Death after Infarction, Attenuating Ventricular Dilation, Wall Stress, and Cardiac Hypertrophy. J Clin Invest. 1997;100:1991–9. DOI: 10.1172/JCI119730 163, 165

[60] Wang Y, Ahmad N, Wani MA, Ashraf M. Hepatocyte growth factor prevents ventricular remodeling and dysfunction in mice via Akt pathway and angiogenesis. J Mol Cell Cardiol. 2004;37:1041–52. DOI: 10.1016/j.yjmcc.2004.09.004 163

[61] Jayasankar V, Woo YJ, Pirolli TJ, Bish LT, Berry MF, Burdick J, et al. Induction of angiogenesis and inhibition of apoptosis by hepatocyte growth factor effectively treats postischemic heart failure. J Card Surg. 2005;20:93–101. DOI: 10.1111/j.0886-0440.2005.200373.x 163

[62] Nakamura T, Mizuno S, Matsumoto K, Sawa Y, Matsuda H, Nakamura T. Myoardial protection from ischemia/reperfusion injury by endogenous and exogenous HGF J Clin Invest. 2000;106:1511–9. DOI: 10.1172/JCI10226 163

[63] Ueda H, Nakamura T, Matsumoto K, Sawa Y, Matsuda H, Nakamura T. A potential cardioprotective role of hepatocyte growth factor in myocardial infarction in rats. Cardiovasc Res. 2001;51:41–50. DOI: 10.1016/S0008-6363(01)00272-3 163

[64] Urbanek K, Rota, M., Cascapera, S., Bearzi, C., Nascimbene, A., De Angelis, A., Hosoda, T., Chimenti, S., Baker, M., Limana, F., Nurzynka, D., Torella, D., Rotatori, F., Rastaldo, R., Musso, E., Quaini, F., Leri, A., Kajstura, J., Quaini, E., Anversa, P. Cardiac Stem Cells Possess Growth Factor-Receptor Systems That After Activation Regenerate the Infarcted Myocardium, Improving Ventricular Function and Long-Term Survival. Circ Res. 2005;97:663–73. DOI: 10.1161/01.RES.0000183733.53101.11 163

[65] Linke A, Muller, P., Nurzynska, D., Casarsa, C., Torella, D., Nasclimbene, A., Castaldo, C., Cascapera, S., Bohm, M., Quaini, F., Urbanek, K., Leri, A., Hintze, T.H., Kajstura, J., Anversa,

P. Stem cells in the dog heart are self-renewing, clonogenic, and multipotent and regenerate infarcted myocardium, improving cardiac function. PNAS. 2005;102:8966–71. DOI: 10.1073/pnas.0502678102 163

[66] Suleiman MS, Singh RJ, Stewart CE. Apoptosis and the cardiac action of insulin-like growth factor I. Pharmacol Ther. 2007;114:278–94. DOI: 10.1016/j.pharmthera.2007.03.001 165

[67] Webster KA. Programmed death as a therapeutic target to reduce myocardial infarction. Trends Pharmacol Sci. 2007;28:492–9. DOI: 10.1016/j.tips.2007.07.004 165

[68] Tomita N, Morishita R, Taniyama Y, Koike H, Aoki M, Shimizu H, et al. Angiogenic property of hepatocyte growth factor is dependent on upregulation of essential transcription factor for angiogenesis, ets-1. Circulation. 2003;107:1411–7. DOI: 10.1161/01.CIR.0000055331.41937.AA 165

[69] Nakamura T, Matsumoto K, Mizuno S, Sawa Y, Matsuda H. Hepatocyte growth factor prevents tissue fibrosis, remodeling, and dysfunction in cardiomyopathic hamster hearts. American journal of physiology. 2005;288:H2131–9. DOI: 10.1152/ajpheart.01239.2003 165

[70] Wang Y, Ahmad N, Wani MA, Ashraf M. Hepatocyte growth factor prevents ventricular remodeling and dysfunction in mice via Akt pathway and angiogenesis. J Mol Cell Cardiol. 2004;37:1041–52. DOI: 10.1016/j.yjmcc.2004.09.004 165, 167

[71] Dobaczewski M, Gonzalez-Quesada C, Frangogiannis NG. The extracellular matrix as a modulator of the inflammatory and reparative response following myocardial infarction. J Mol Cell Cardiol. 2010;48:504–11. DOI: 10.1016/j.yjmcc.2009.07.015 165

[72] Frantz S, Bauersachs J, Ertl G. Post-infarct remodelling: contribution of wound healing and inflammation. Cardiovasc Res. 2009;81:474–81. DOI: 10.1093/cvr/cvn292 165

[73] Shriver Z, Liu D, Sasisekharan R. Emerging views of heparan sulfate glycosaminoglycan structure/activity relationships modulating dynamic biological functions. Trends Cardiovasc Med. 2002;12:71–7. DOI: 10.1016/S1050-1738(01)00150-5 165

[74] Haider H, Akbar SA, Ashraf M. Angiomyogenesis for myocardial repair. Antioxidants & redox signaling. 2009;11:1929–44. DOI: 10.1089/ars.2009.2471 167

[75] Pasumarthi KB, Field LJ. Cardiomyocyte cell cycle regulation. Circ Res. 2002;90:1044–54. DOI: 10.1161/01.RES.0000020201.44772.67 169

[76] Jopling C, Sleep E, Raya M, Marti M, Raya A, Belmonte JC. Zebrafish heart regeneration occurs by cardiomyocyte dedifferentiation and proliferation. Nature. 2010;464:606–9. DOI: 10.1038/nature08899 169

[77] Rysa J, Tenhunen O, Serpi R, Soini Y, Nemer M, Leskinen H, et al. GATA-4 is an angiogenic survival factor of the infarcted heart. Circ Heart Fail. 2010;3:440–50.
DOI: 10.1161/CIRCHEARTFAILURE.109.889642 169

[78] Pikkarainen S, Tokola H, Kerkela R, Ruskoaho H. GATA transcription factors in the developing and adult heart. Cardiovasc Res. 2004;63:196–207. DOI: 10.1016/j.cardiores.2004.03.025 169

[79] Urbanek K, Rota M, Cascapera S, Bearzi C, Nascimbene A, De Angelis A, et al. Cardiac stem cells possess growth factor-receptor systems that after activation regenerate the infarcted myocardium, improving ventricular function and long-term survival. Circ Res. 2005;97:663–73. DOI: 10.1161/01.RES.0000183733.53101.11 169

[80] Landa N, Miller L, Feinberg MS, Holbova R, Shachar M, Freeman I, et al. Effect of injectable alginate implant on cardiac remodeling and function after recent and old infarcts in rat. Circulation. 2008;117:1388–96. DOI: 10.1161/CIRCULATIONAHA.107.727420 172

[81] Leor J, Tuvia S, Guetta V, Manczur F, Castel D, Willenz U, et al. Intracoronary injection of in situ forming alginate hydrogel reverses left ventricular remodeling after myocardial infarction in Swine. J Am Coll Cardiol. 2009;54:1014–23. DOI: 10.1016/j.jacc.2009.06.010 172

Authors' Biographies

EMIL RUVINOV

Emil Ruvinov obtained his Ph.D. in Biotechnology Engineering in 2011 from the Ben-Gurion University of the Negev, Israel, where he is now a post-doctoral researcher at the Cohen group. His main area of research is biomaterial-based drug delivery for tissue engineering and regenerative medicine applications. He has published several papers and co-authored several book chapters in the field of cardiac tissue engineering and regeneration.

YULIA SAPIR

Yulia Sapir obtained her B.Sc. (2008, Suma Cum Laude) and M.Sc. (2010, Suma Cum Laude) in Biotechnology Engineering from the Ben-Gurion University of the Negev, Israel, where she is now a Ph.D. candidate (Azrieli Fellow) in Prof. Smadar Cohen laboratory. Her main research interests is engineering of functional cardiac patch, focusing on innovative biomaterial design and investigation of different stimulation patterns.

SMADAR COHEN

Prof. Smadar Cohen is the Claire and Harold Oshry Professor in Biotechnology and co-founder of the Avram and Stella Goldstein-Goren Department of Biotechnology Engineering at Ben-Gurion University of the Negev, Beer Sheva, Israel. Her main research interests are in the design of bio-inspired materials and advanced bioreactors for tissue engineering and regeneration *as well as* implementation of intelligent nano-sized delivery systems for therapeutics. Professor Cohen published over 150 publications, edited two books and is the inventor of over 35 US patents.

Printed in the United States
by Baker & Taylor Publisher Services